CURRENT STEM

CURRENT STEM

VOLUME 2

CURRENT STEM

MAURICE HT LING (SERIES EDITOR)
PRINCIPAL PARTNER, COLOSSUS TECHNOLOGIES LLP, SINGAPORE

Current STEM. Volume 2
Maurice HT Ling (Editor)
2019. ISBN: 978-1-53616-042-0
(softcover)
2019. ISBN: 978-1-53616-043-7
(eBook)

Current STEM. Volume 1
Maurice HT Ling (Editor)
2018. ISBN: 978-1-53613-416-2
(hardcover)
2018. ISBN: 978-1-53613-417-9
(eBook)

CURRENT STEM

VOLUME 2

MAURICE HT LING
EDITOR

Copyright © 2019 by Nova Science Publishers, Inc.

All rights reserved. No part of this book may be reproduced, stored in a retrieval system or transmitted in any form or by any means: electronic, electrostatic, magnetic, tape, mechanical photocopying, recording or otherwise without the written permission of the Publisher.

We have partnered with Copyright Clearance Center to make it easy for you to obtain permissions to reuse content from this publication. Simply navigate to this publication's page on Nova's website and locate the "Get Permission" button below the title description. This button is linked directly to the title's permission page on copyright.com. Alternatively, you can visit copyright.com and search by title, ISBN, or ISSN.

For further questions about using the service on copyright.com, please contact:
Copyright Clearance Center
Phone: +1-(978) 750-8400   Fax: +1-(978) 750-4470   E-mail: info@copyright.com.

NOTICE TO THE READER

The Publisher has taken reasonable care in the preparation of this book, but makes no expressed or implied warranty of any kind and assumes no responsibility for any errors or omissions. No liability is assumed for incidental or consequential damages in connection with or arising out of information contained in this book. The Publisher shall not be liable for any special, consequential, or exemplary damages resulting, in whole or in part, from the readers' use of, or reliance upon, this material. Any parts of this book based on government reports are so indicated and copyright is claimed for those parts to the extent applicable to compilations of such works.

Independent verification should be sought for any data, advice or recommendations contained in this book. In addition, no responsibility is assumed by the Publisher for any injury and/or damage to persons or property arising from any methods, products, instructions, ideas or otherwise contained in this publication.

This publication is designed to provide accurate and authoritative information with regard to the subject matter covered herein. It is sold with the clear understanding that the Publisher is not engaged in rendering legal or any other professional services. If legal or any other expert assistance is required, the services of a competent person should be sought. FROM A DECLARATION OF PARTICIPANTS JOINTLY ADOPTED BY A COMMITTEE OF THE AMERICAN BAR ASSOCIATION AND A COMMITTEE OF PUBLISHERS.

Additional color graphics may be available in the e-book version of this book.

Library of Congress Cataloging-in-Publication Data

ISBN: 978-1-53616-042-0

Published by Nova Science Publishers, Inc. † New York

CONTENTS

Preface		vii
Chapter 1	Perils and Pitfalls of Idiomatic Python *K. S. Ooi*	1
Chapter 2	A Reflection of My 4 Years of Undergraduate Study in Singapore: What I Learnt from Singapore *Argho Maitra*	45
Chapter 3	SEcured REcorder BOx (SEREBO) Version 1.0 *Maurice HT Ling*	67
Chapter 4	The Changes While Studying in the Different Countries and Reflections to the Moments *Gyun Tae Bae*	155
Chapter 5	My Memoir Based on 4 Years in Singapore *Jung Hwan Kim*	167
About the Editor		177
Series Description		179
Index		183
Related Nova Publications		187

PREFACE

Science, technology, engineering, and mathematics; collectively acronym as STEM; are rapidly advancing fields in their own rights. As a book series, Current STEM aims to be a friendly forum for both academic researchers and industrial practitioners to present their work as book chapters. Hence, the chapters should be varied, and this is intended. Current STEM is encompassing in terms of the type of work to encourage a generation of researcher-practitioners. In this volume, there are three personal narratives of personal experiences from overseas students.

I hope this second volume will be a pleasure to a wide audience.

Maurice HT Ling, PhD

Chapter 1

PERILS AND PITFALLS OF IDIOMATIC PYTHON

K. S. Ooi[*]
Faculty of Information Technology and Science,
INTI International University, Persiaran Perdana BBN,
Putra Nilai, Negeri Sembilan, Malaysia

ABSTRACT

Learning to program in Python is difficult. The use of idiomatic Python adds a layer of complexity in coding. Python programmers who want to impress have such a strong urge to be fashionable that they are too eager to use idiomatic Python in their code. This blind conviction more often than not causes these programmers to produce convoluted and unreadable code, especially when the goalposts or requirements of the problems are constantly moving. In this chapter, the author shows that unrestrained use of idiomatic Python may not be that suitable even in some simple problems with dynamic requirements. The author uses examples to support and illuminate the claim that Python programmers should focus on problem solving rather than on the use of idiomatic Python. Even though idiomatic Python is a hype, a good Python programmer had better be prepared to be verbose to solve a problem, and

---
[*] Corresponding Author's E-mail: kuansan.ooi@newinti.edu.my.

only use idiomatic Python when its expressiveness does not convolute the code.

Keywords: python 3, idiomatic python, string formatter, idiomatic swap, enumerate, list comprehension, reversed, lambda, slicing, exceptions, generator expressions

INTRODUCTION

Imagine walking into a bookstore looking for books that could help you learn a programming language, which in our case is Python, you would be led to think that coding is not a big deal. Quite a number of books seem to suggest that you can teach yourself Python in "24 hours." That can be misleading, and a real software project can expose how little knowledge acquired from such sources can truly be applied. As a beginner, you can start writing some Python scripts in about 20 learning hours (Oldham, 2005; Orfanakis and Papadakis, 2016). This estimation is not without a basis, if we buy Kaufman's explanation (Kaufman, 2014). But learning to code in Python with an acceptable degree of fluency requires time, and evidence seems to point to about 10,000 hours (Gladwell, 2009; Norvig, 2014). Though 10,000 may seem like a trifling sum, it connotes hard work and dedication. This is especially true when you want to be trendy in Python coding, that is you want to code like a Pythonista or coder extraordinaire who uses a lot of idiomatic Python. Coding is fundamentally a human activity (Weinberg, 1971), and therefore pride may be central (Tracy, 2016). This chapter is about some aspects of idiomatic Python that may result in unreadable code due to pride, but the chapter does not address some grand issues of psychology of programming directly.

Can we blame the Python programmers who want to impress, after all Python gives us so many ways to write cool code? Let us take a step back and talk about idioms in English. An idiom is a phrase that is often difficult to comprehend by looking at its individual words, for example "a hot potato." When we become well versed in English, we tend to infuse our speech with idioms to add color and flavor so that we can demonstrate to any English speaking audience our command over the language. We should do this as a natural progression as we use English more and more daily. However, the overuse of idioms in a speech may give the impression of a showing off, rather than communicating. Likewise, if we use idiomatic Python simply because we want to impress others with our mastery, or to obfuscate the code egoistically to garner admiration, then be rest assured that others are going to find difficulties in reading our code. Yet, when we have used enough Python, just like after we have become well versed in English, idioms tend to slip into our daily usage naturally without us thinking about the underlying fundamentals. Understanding idiomatic Python is crucial, if we want to know why we do not use it sometimes. Partial understanding leads to misconceptions, which are particularly pernicious.

I may come across as someone with a negative perception towards idiomatic Python and for that I will like to expatiate a bit more on some of the points I made in the preceding paragraph. The use of idioms is inevitable. Soon or later, we will sprinkle some of the idioms everywhere in our code, but we should do that solely to simplify the code and make it more readable, and not to show off our mastery of idioms. If we insist on using idioms consciously to make our code look more impressive, then rest assured that our code would suffer from poor readability. The code produced to impress using idioms will be difficult to read by others or by yourself sometime in the future, let alone maintained. In short, the use of idiomatic Python in moderation with solid understanding will bring clarity to our code, and put us in right standing among the Python programmers.

Each of the Python code found in this chapter is code snippet, and hence you may need to add more code to get it interpreted or compiled. I used Python 3.5 to compile all the snippets. If you are interested in porting the code to Python 2, please refer the book by Nanjekye (Nanjekye, 2017). If you subscribe to classifying computer programs to levels, see for example (Skrien, 2009), you will find that the code of this chapter either tries to find a better way to write the basic structures of Python or look for a better coding practice to perform tasks. There may be some attempts to find efficient ways, but mostly this chapter endeavors to present what the author thinks can be a candidate of a cleaner and more manageable Python code. The author has not undertaken any attempt to design a better system using object-oriented design or functional programming. These levels of programming will be reserved for future writings. Furthermore, whatever lies ahead is rather opinionated. You should not agree to all of them.

Table 1. Selected variable names used in this chapter

Variable names	Explanation
`i`, `j`, `k`, and `p`	Index variables
`a`, `b`, and `c`	Integer variables for swapping
`a`	A list of integers
`x`	An item in a list
`n`	Size of `a`, that is `len(a)`. Size of `s` as well.
`na`	Integer division of `n` by 2, that is `n//2`
`s`	A string variable
`swap`	Temporarily swapping variable
`g`	Generator object

Before moving on, you may ask where I get the healthy dosage of idiomatic Python to condition myself to write the code. The book you should read to bask yourself in is Knupp's (Knupp, 2013). There are a number of websites writing about idiomatic Python, but if you insist on reading one, visit Goodger's (Goodger, 2008). Kaleniuk (Kaleniuk, 2017) expresses his opposition openly to programmers who do not question the didactic style of Knupp's (Knupp, 2013); this chapter, in contrast, does not attempt to attack any source but promotes the use of idiomatic Python with hindsight. If you have time to watch only one YouTube video, then watch Hettinger's (Hettinger, 2013). Books that stand out in their coverage on idioms are Slatkin's (Slatkin, 2015), Ramalho's (Ramalho, 2015), Reitz, and Schlusser's (Reitz, and Schlusser, 2016), and Jaworski and Ziadé's (Jaworski and Ziadé, 2016). Overall, it is often difficult to say what is idiomatic and what is not; opinions are divided. That does not stop me from writing this chapter. I will digress first with the Zen and then start you off with something rather trivial. (I will use the character ↳ to break code into multiple lines if such break improves the presentation.)

INTERLUDE: THE ZEN OF PYTHON

It is hard to pinpoint exactly the domain of idiomatic Python without looking at opinionated writings I used above. Let us try to take cue from PEP 20 (Peters, 2004; Warsaw, 2010), *The Zen of Python*. Almost all learners of Python programming I have ever met so far know about the Zen. But the level of knowledge is usually minimal, often as minimal or limited as knowing the name "Zen." I have not met a Zen cognoscenti yet. The noble aim of the Zen, according to the official PEP 20 site (Peters, 2004), is to "succinctly [channel] the BDFL's guiding principles for Python's design into 20 aphorisms." However, only 19 of the 20 aphorisms were actually written down. This does not stop many Python developers from using the Zen as a guide to their coding. A book by Reitz and Schlusser (Reitz and Schlusser, 2016), a presentation by Blanks (Blanks, 2011), and a talk by Hettinger (Hettinger, 2013) demonstrate how to apply the Zen's aphorism in Python coding.

Whenever I read the Zen, it never fails to impress upon me that Tim Peters is talking to me by the iceberg theory of communication (Krogerus and Tschäppeler, 2018), intended or unintended, with the deeper meaning of the Zen hidden underneath. Aphorism is a typical example of iceberg way of saying things. I personally like to read aphorisms, but I am not a fan of people who use them to make decision, particularly in computer code. Aphorisms are pearls of wisdom, but they are open-ended, contradictory, tautological, and so forth. For example, one of the aphorisms from Haspel's book (Haspel, 2015) about lying is "To be better it is first necessary to pretend to be; and objections to improvement often masquerade as objections to pretense." Yes, it sounds smart, but I have difficulty to dive below the surface of the iceberg to get exactly what the author wants to impart. On the contrary, Peterson (Peterson, 2018) devotes a whole chapter on why we should not lie with discourse, antidotes and arguments before reaching the conclusion "Tell the truth. Or, at least, don't lie." In the end, I learn something concrete from Peterson's but not from Haspel's.

Let us get back to PEP 20. There are times the aphorisms offer very little help if we do not get the context the code is written. For example, the Zen "explicit is better than implicit" has difficulty labeling the Bad from the Good in the following snippets, because the rationale of using the pathlib is unclear.

```
# (1)
import os
if os.path.isfile(file_name):
    os.remove(file_name)

# (2)
import pathlib
p = pathlib.Path(file_name2)
p.touch()
if p.exists():
    p.unlink()
```

Let us take another little "swipe" on Zen 6 "*sparse is better than dense*," using an example I have created but inspired by an example given in Reitz and Schlusser (Reitz and Schlusser, 2016). Out of the following three snippets, the first one is the densest, but at the same time the clearest. I arbitrarily asked some colleagues around the office to decide which snippet is their preference without explaining the "why" of their choice; I found out that all of them like the first snippet. It was as if a *blink* machine inside that help them to make the snap judgment (Gladwell, 2007).

```
# (1)
if 2 <= x <= 3 or 50 <= x <= 60:
  # do something ...

# (2)
if x >= 2 and x <= 3 or x >= 50 and x <= 60:
```

```
    # do something ...

# (3)
cond1 = x >= 2 and x <= 3
cond2 = x >= 50 and x <= 60
if cond1 or cond2:
    # do something ...
```

We cannot deny that frame or narrative context (Krogerus and Tschäppeler, 2018) is an important tool in convincing others to reject or accept an idea. Tim Peters's Zen does give us an effective framing tool when we want to take swipe at, say, the ORM design pattern of SQLAlchemy (Banks, 2011). We cannot deny the usefulness of the Zen as a language tool when we want to argue with someone about the Good or Bad of a piece of code. Still, you will not get consensus in toto. Using the preceding example, if you want to argue for the third snippet, you can use the sparse and dense and even back it up with Peters's credentials, but I will still say that the first is the clearest and I bet many Python coders would be in my camp. On the other hand, I would love to imagine myself using the Zen to argue against the ORM, just as Banks did. I would still have hard time to sway even one SQLAlchemy's user. To have a better success rate of swaying, I would need Peterson's story-telling skill, and not a tirade of aphorism.

Next, let us explore the relationship between the Zen and idioms. According to Reitz and Schlusser (Reitz and Schlusser, 2016), the particular aphorisms that subsume idioms are Zen 13 and 14, which are "*There should be one – and preferably only one – obvious way to do it,*" and "*Although that way may not be obvious at first unless you're Dutch,*" respectively. The Zen 13 *seems* to propagate that the idiomatic code is the way of Python. The Zen 14 *seems* to indicate that good idiomatic code must be acquired, and it may not be obvious at first to beginners (I will skip the Dutch here). In some circumstances, Zen 13 transcends, or rather idiom transcends. For example, if you want to find out whether a sentence s has a word "way" in it using a print statement,

```
# (1)
print('way' in s)

# (2)
t = s.split(' ')
for w in t:
    if 'way' == w:
        print(True)
```

it would be foolish if one opts for the second snippet. However, in many other circumstances, *the one and only one obvious way to do thing with Python* does not exist, as we will see in this chapter. There may exist multiple good ways of writing a code. We will show you that if you insist *strictly* on jumble up some so-called idiomatic pieces to build a larger whole, you are bound to end up having an unreadable code. The Zen seems to suggest that there is a platonic Python code for solving a problem and if you believe there is one, you will find it, but others may not concur that what you have found is the one.

Do not forget that code writing is fundamentally a human activity and pride may be the driving force. We are different people, and we have different taste. The Zen may transcend, as I have shown you; it may also be a bad decider, as I have also shown you in the sparse and dense example. Furthermore, it is immensely difficult to determine what is the one and only way to write idiomatic Python code using the Zen as a guide, because the aphorisms are too susceptible to misunderstanding. One can easily create a corpus of counterexamples for each Zen, but this is not the intent of this chapter. Since good idioms are the precipitation of the Python programming, and the adoption of JSON en masse has not killed off SQLAlchemy (SQLalchemy, 2018), I will steer clear of the Zen in this chapter and focus instead on sources that propagate *directly* what a piece of idiomatic code should be. That also means that while I am steering clear of aphorism of Zen, I will not be able to avoid words used in Zen, such as readability, when I argue my case. I will try to steer these words away from the Zen, and use them as they are used by Maguire (Maguire, 1993) and Martin (Martin, 2008).

PEP 8 AND STRING FORMATTER

After the digression, let us start this chapter with two good stuffs of idiomatic Python: a little coding standard advocated by PEP 8 (van Rossum et al., 2001), and the use of *string formatter* to construct strings. Python community endorses the use of PEP 8 coding convention, particularly when you create open-sourced software packages. Complying with this convention in your coding is first and foremost Pythonic. When you are using this coding convention subconsciously and spontaneous, you have achieved some level of idiomatic mastery. PEP 8 is a good thing. However, few latest IDEs I know do not enforce nor recommend; thus, getting your code compiled with them does not guarantee that you have followed PEP 8. Sadly, to streamline my presentation, I am going to favor the use of the following naming of selected variables used in this chapter, which are given in the following table.

Since I also program in other languages and have done some reading, I came across a book (Liguori and Liguori, 2017) with a passage that read: "Temporary variable names may be single letters such as `i`, `j`, `k`, `m`, and `n` for integers and `c`, `d`, and `e` for characters." Even though I try to sneak that in to justify the choices of variable names I use in this chapter, that does not mean I have forsaken PEP 8. Again, I have to reiterate that in real software development, good names that complied with PEP 8 are crucial. You do not have to read the PEP 8 in full at (van Rossum et al., 2001) to resist the use of names such as `an_array_of_numbers`, `the_index_of_array`, etc.

An idiomatic Python that I wholeheartedly support is the use of *string formatter* to construct strings. Let me give you an example.

```
s = "{} is {} m tall, and {} weighs {} kg."
print(s.format("Dwayne Johnson", 1.96, "he", 118))
print(s.format("Vin Diesel", 1.85, "he", 102))
print(s.format("Scarlett Johansson", 1.60, "she", 57))
```

Three new strings are constructed from the string object s, by passing three different format messages to the formatter. This is rather straightforward stuff - that is, as long as you know that you are using objects to construct these three strings, you have achieved some satisfactory understanding of the code. Despite its simplicity, string formatter is quite useful, which I will discuss further before I end this section below.

However, I dislike somewhat the use of *format operator* % to construct strings. Let me rewrite the above snippet using this operator.

```
s = "%s is %g m tall, and %s weighs %d kg."
print(s % ("Dwayne Johnson", 1.96, "he", 118))
print(s % ("Vin Diesel", 1.85, "he", 102))
print(s % ("Scarlett Johansson", 1.60, "she", 57))
```

Why so? First, it reminds us of the conversion specification of the control string to printf and sprintf functions from C, with even more % character. Unfortunately, it has its cool factor, as I have seen its rampant use by Python programmers. Second, it uses the same symbol as the reminder operator %, as if we have not had enough of it. However, when I use the operator in few occasions, it turns out that it is as powerful as the string formatter, and it is hard to tell how soon in the near future I will stop complaining.

The curly braces {} in s of string formatter is the placeholder at which you substitute it with some value to construct the resulting string. I can also place indices into the placeholders. The following snippet illustrates the use of indices in string formatter.

```
print("I like {0} and {1} but not {2}. "
    ↳.format("Python","R","Java"))
print("I like {2} and {1} but not {0}. "
    ↳.format("Python","R","Java"))
```

The printouts from these two print statements should be obvious to you. You can also insert keywords into curly brace when constructing a string. For example:

```
print("The radius of the top circle is {top} cm while the
radius of the bottom circle is {bot} cm ."
↳.format(top = 7.2, bot = 5.7))
```

You can use string formatter to display a floating number by adding the :f argument into the placeholder, as shown below.

```
print("Adrian      scored      {0:f}      for      his
{1}!".format(90.75,"Physics"))
```

The output above generates many zeros after the decimal point. To limit the decimal places of the number, you can, for example, use {0:.2f} argument instead.

On the other hand, the *format operator* % is as powerful as string formatter. Let us look at the following examples.

```
# Display string,%s
print("My sister like to eat %s, %s, and %s. " %
   ↳("orange","apple","watermelon"))

# Display integer,%d
print("Keith buy %d eggs, %d burger and %d chocolate cakes
from the market." % (5,1,2))

# Display float,%f
print("I got %.2f for my math quiz." % (70.23))
```

Each time I want to use format operator %, I have to specify the type of object that I am going to replace. For instance, s is for strings, d is for integers and f is for floating point numbers. If I give the wrong type of object that I am going to replace, I will get an error.

SWAPPING

Swapping values without the use of temporary variable is idiomatic. If that is how much you know, I suggest that you stay away from idiomatic swapping.

If you want to swap values held by variable a and b, the pundits will beckon you to write

```
a, b = b, a
```

instead of the swap that uses a temporary variable

```
swap = a
a = b
b = swap
```

The pundits would say the use of temporary variable is a Python backwater or backwoods. Regardless of what the pundits say, you must understand the data structure behind the idiomatic swap or risk producing code that you will regret showing to others.

Say you were asked to swap three variables: a to b, b to c, and c to a. If you wrote

```
a, b = b, a
b, c = c, b
```

then you had misunderstood idiomatic swap. The solution is correct, but silly, to say the least. The better solution, if you understand that the idiomatic swap actually involves the creation of tuples, is

```
a, b, c = b, c, a
```

Tuples are immutable data structures that keep values. Typically, we use tuples to keep key values in dictionaries, another commonly used data structure in Python. Here, we use tuples to swap values. If you learn the idiomatic swap without knowing tuple, you will have no idea that the better solution actually involves the creation of two tuples, one on the left and the other on the right of the assignment operator. The silly solution before this involves the creation of four tuples. Python programmers have the habit of leaving out the enclosing parentheses when dealing with tuples. If you put back the parentheses to the better solution, the tuples become proper.

```
(a, b, c) = (b, c, a)
```

Leaving parentheses out in tuples is not an issue here; the choice is yours. Creation of these two tuples require resources, and, hence, you should not expect that the use of idiomatic swap would give you a meteoric rise to performance over the use of temporary variable. Let us use the selection sort, which is hungry for swapping during sorting, to perform an experiment. First,
I created random numbers using the Sattolo's algorithm (Sattolo, 1986) and stored them in a file data.txt.

```
from random import *
a = list(range(100000))
n = len(a)
i = n
while(i > 1):
        r = int((i - 1)*random())
        temp = a[r]
        a[r] = a[i - 1]
        a[i - 1] = temp
        i = i - 1
file = open("data.txt", "w")
for i in range(n):
        file.write(str(a[i]) + '\n')
file.close()
```

Then, I ran the following two selection sorts to sort the random numbers, and recorded the time taken by each version to sort these numbers, following the advice from (Gorelick and Ozsvald, 2014).

```
# (1) using idiomatic swap (idiomatic)
for i in range(0, n - 1):
    p = i
    for j in range(i + 1, n):
        if(a[j] < a[p]):
            p = j
    if(p != i):
        a[p], a[i] = a[i], a[p]


# (2) using a temporary variable for swapping (Temp-Var Swap)
for i in range(0, n - 1):
    p = i
    for j in range(i + 1, n):
        if(a[j] < a[p]):
            p = j
    if(p != i):
        swap = a[p]
        a[p] = a[i]
        a[i] = swap
```

This is a traditional way to measure performance, and I expect many pundits to disagree. That aside, what I want to see is *performance boost* from the use of idiomatic swap. After repeating the experiment many times, I found no appreciable performance boost from either version. When I say I repeated the experiment, what I mean is that each of the two sorting versions was run 20 consecutive times to sort 5,000 to 50,000 random numbers. The result is shown in Figure 1. The Python script that was used in the experiment is listed in Appendix B. One important realization is that the complexity of algorithm in both sorting versions does not change, that is both take $O(n^2)$ time (Aho, Hopcroft, and Ullman, 1983). Swapping for selection sort takes a constant time, whether the swap is idiomatic or using a temporary variable, but when the sorting size increases, we expect the running time of selection sort algorithm to be $O(n^2)$. However, optimization of Python code is still a preoccupation for some (Gorelick and Ozsvald, 2014), and making your code run faster in a particular machine may pay dividend. The optimization path may lead you as far as Cython (Cython.org, 2018); this path is interesting, and by itself can be a preoccupation. As a reminder, wrong choice of algorithm is the main cause of your slow programs. It is an open secret that if you use the selection sort, either version, to sort, say, 200 million random numbers, the sorting will grind to a halt (Aho, Hopcroft, and Ullman, 1983). Quicksort, for example, will sort the numbers in a reasonable time.

The upshot is unless you know tuples well, I suggest that you shun the use of idiomatic swap. Furthermore, focus on the correct algorithm instead of the correct idiomatic code. You will be rewarded.

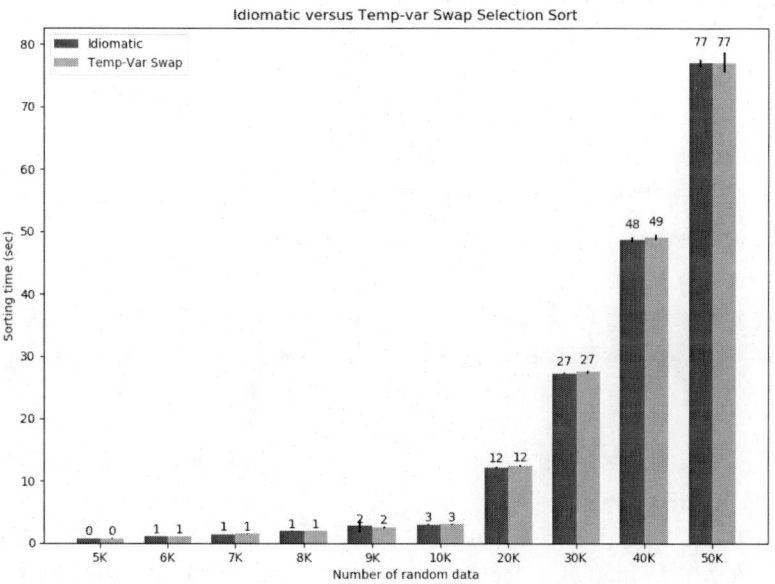

Figure 1. The figure shows that no significant performance boost from using idiomatic swap in selection sort.

ENUMERATION AND LIST COMPREHENSION

Give any Python programmer enthusiast full rein, insofar as he or she always wants to be trendy, the code that this individual produces is a statement of his or her campaign against verbosity. He or she will fervently embrace idioms. Let us call this programmer a Pythonist, instead of

Pythonista. If a Pythonist insists on learning the idioms per se and ignore the fundamentals, this individual forgets that the code is for others to read. This individual also limits the problems that he or she can tackle effectively. The problem lies in our bias in solving problems from the viewpoint of Déformation Professionnelle (Dobelli, 2013). The Pythonist simply could not see other better ways of writing code except through idioms.

Let us look at a common thing we often do as a Python programmer: visiting each element of a list, and along the way making an observation on or modification to each element. Idiomatic Python has tentacles to this chore. Let us start with a seemingly simple problem of finding the smallest number from a list. If you want to get an approximate smallest number, there are numerous creative answers. However, if the exact smallest number is what you seek, your code has to examine each element in the list. A verbose way in which you can do this is given below, where a is the list of numbers and n is `len(a)`.

```
# (1a) Using for-range (verbose)
smallest = a[0]
for i in range(1, n):
    if a[i] < smallest:
        smallest = a[i]
```

You will immediately get into the nerve of a Pythonist. The Pythonist would say enough is enough and suggest that you do

```
# (2a) Using min(a)
smallest = min(a)
```

I performed an experiment similar to that of selection sort. For finding that smallest number among 100,000 random numbers, the verbose solution and the idiomatic solution were both blazingly fast (see Appendix C). If you had a little time to invest, you would find that Python is implemented in c, and you expect the built-in function min would be faster when the size of the array increases. It comes with no surprise that the verbose way of locating the minimum with a million random numbers is within the order of hundredth of seconds slower in my machine (see Appendix C). I am not too concern about this; obviously, if you want to find the minimum per se, the idiomatic way should be the one, after all who wants to write code that finds the minimum of an array when a built-in function is available. You will soon find out that this is not what matters us most. Let us put this thread of thought aside, and add one more requirement in addition to finding that smallest number. For instance, if you are required to find the index for this smallest number, the verbose solution, with an additional variable, may look like this:

```
# (1b)
smallest = a[0]
smallest_index = 0
for i in range(1, n):
        if a[i] < smallest:
                smallest = a[i]
                smallest_index = i
```

If you were a Pythonist, you would not fail to recall that *enumerate* sugar. Armed with the impeccable knowledge of tuple, you might modify the verbose solution as follows:

```
# (2b)
smallest, smallest_index = a[0], 0
for i, item in enumerate(a):
        if item < smallest:
                smallest, smallest_index = item, i
```

This solution runs slower compared to the verbose one. Furthermore, the loop visits the list one extra time, because enumerate start with the index 0. A naïve Pythonist who is misconceived will try to set the *start* parameter to 1:

```
# (3b) A wrong algorithm!
for i, item in enumerate(a, start = 1):
```

If you read the documentation, the start parameter that is set to 1 does not direct the loop to start with the second element of the list, and this much I want to say here. Since enumerate returns tuples, some smart Pythonists would suggest that you cast the tuple to a list object and has it start with the second element:

```
# (4b)
for i, item in list(enumerate(a))[1:]:
```

Unfortunately, this solution worsens the performance. I performed an experiment on my machine by modifying the script shown in Appendix B. These three algorithms (1b, 2b, and 4b) are subjected to finding the minimum and its index of array of random numbers. I repeated the experiment using different array sizes varying from 100,000 to 1,000,000. The result is shown in Figure 2, which clearly shows that (4b) performs the poorest among the three algorithms, and (1b) slightly outperforms (2b). This is an example that *enumerate* does not always yield a better solution than *range*. Since I am conditioned to write idiomatic Python before writing this chapter, I told myself *wait a minute*, just like a Pythonist would say, how about the following solution, a cool solution!

```
# (4e)
smallest = min(a)
smallest_index = a.index(smallest)
```

I have to admit that the solution is elegant. However, I still want to argue that a Pythonist can be coerced into producing bad code if he or she ignores the fundamentals of algorithm and requirements, and insists that idiomatic Python is the way to produce code, a Déformation Professionnelle bias I referred to earlier. Let us continue with the example, upon finding the smallest element, for instance, you are required to put the smallest element into the head of the list. I can imagine that the Pythonist in me is happy to supply the following line as an additional code after the previous one.

```
a[smallest_index], a[0] = a[0], a[smallest_index]
```

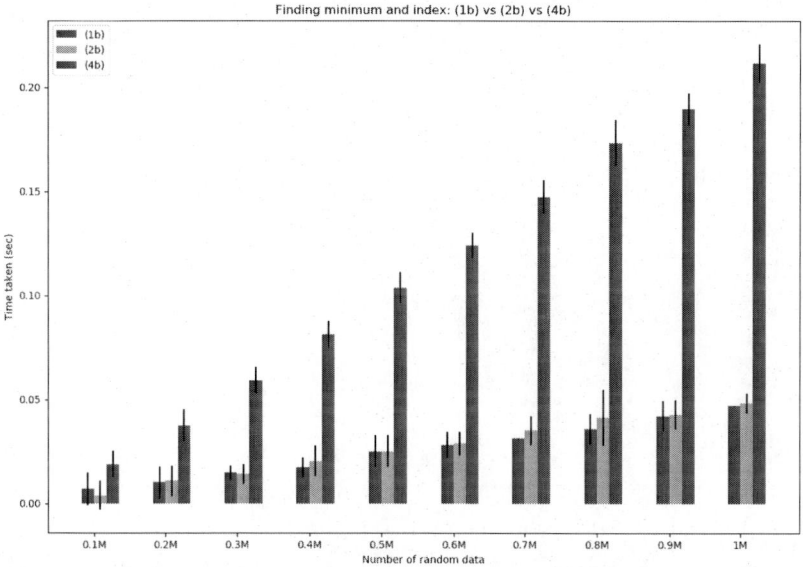

Figure 2. Comparing performance of (1b), (2b), and (4b) for finding minimum and its index of array of random number. The array sizes vary from 100,000 to 1,000,000.

A cool solution again, with just one additional line of code. Let say the requirement changes, because your manager realizes that there may be more than one smallest element in the list, and you are required to move all the same smallest elements into the front of the list, not just the first occurrence of the smallest element. Just sort it!

```
a.sort()
```

Just when you think you have nailed it, the requirement goes up yet another notch! This time you are asked not to disturb the positions of other elements in the list, as much as possible. Back to the drawing board, the enumerate function comes back with a vengeance:

```
smallest = min(a)
head_index = 0
for i, x in enumerate(a):
    if x == smallest:
        a[i], a[head_index] = a[head_index], a[i]
```

```
                head_index = head_index + 1
```

However, list comprehension keeps plaguing the mind. You have to transform the preceding snippet into something like this:

```
smallest = min(a)
for k, p in enumerate(
   ↳[i for i, item in enumerate(a) if item == smallest]):
       a[k], a[p] = a[p], a[k]
```

This code produces the same result and makes no difference in performance as the previous one, but this code is so much incomprehensible, after reducing the number of line of code to one half. I do not think such elegance is justified. The code starts to look obfuscated. Mixing list comprehension and enumerate may not be a good idea after all. Let us give mixing list comprehension and range a shot. The Pythonist in me produces the following code:

```
smallest = min(a)
j = 0
for k in [i for i in range(len(a)) if a[i] == smallest]:
      a[k], a[j] = a[j], a[k]
      j = j + 1
```

This code is equally unreadable. You are not convinced! Show this code to any Pythonist and ask, "What does it do?" A small number of Pythonic high priests might be able to read it offhand, but the rest of us will have hard time deciphering it. We should take the initiative to improve the readability of the solution.

My insistence of using idiomatic Python ends up in producing code that is difficult to understand. Let us get back to basics and forget about enumerate, range and list comprehension, and express the solution of the problem as clearly and readable as possible. Following this urgency, I produce:

```
smallest = min(a)
```

```
i = 0
while(smallest in a[i:]):
    j = a[i:].index(smallest) + i
    a[i], a[j] = a[j], a[i]
    i = i + 1
```

The code is much more readable, for each line is rather short for your brain to digest. You can do better if you proceed, without the insistent or beckoning of idiomatic Python. Before proceeding, let me remind you that the problem we tried to solve was not created in a vacuum; it has real applications. For example, we can use it as the algorithm for an automaton that moves non-performing managers to branches near the headquarters, with minimum impact on the movement of other managers.

Before leaving this section, imagine someone with sound programming fundamentals walked into my office, and said: "May I suggest that we use a massive parallel system to do this job? Wait, give me some stacks, and see what I can do to solve the problem." I would pay much attention to such individual than someone who started the sentence with "The idiomatic solution is …"

REVERSED

Reversing a list is another chore, and the tool wielded by Pythonists is *reversed*. Combined with list comprehension, reversing a list a looks like this:

```
# reversed 1(a)
[x for x in reversed(a)]
```

However, reversing tuples, for example `[(0, 3), (1, 9)]` to become `[(1, 9), (0, 3)]`, using the following snippet is just dead wrong:

```
# reversed 2(a)
```

```
[(i, x) for i, x in enumerate(reversed(a))]
```

This is a common misconception about the use of enumerate. How many of us actually take quality time to read about enumerate function carefully? If you do, you will find out why the above snippet does not produce the correct reversed list. Undeterred and aided by the *zip* function, one of the ways to do it is by using reversed function twice:

```
# reversed 2(b) where n = len(a)
[(i, x) for i, x in zip(reversed(range(n)), reversed(a))]
```

This is the solution proposed by countless pundits to solve the tuple-reversing problem.

I immediately develop an aversion just by looking at this code [reversed 2(b)]. The code simply carries too many Python functions. A bit more of analysis should have you realized that the zip function can be passed over as a parameter, and thus I rewrite the code as follows:

```
# reversed 2(c) where n = len(a)
[(n - i - 1,x) for i, x in enumerate(reversed(a))]
```

If you step away from list comprehension, you can use array slicing (we will discuss slicing later).

```
# reversed 2(d)
list(enumerate(a))[::-1]
```

Or, equivalently you can use the *slice* function

```
# reversed 2(e) where n = len(a)
list(enumerate(a))[slice(-1, -(n + 1), -1)]
```

This way of reversing tuples is quite abstract, and the attempt to hide complexity is a bit over the top. You have to master the use of list function, in addition to enumerate and perhaps slice function. What I am emphasizing is that you really do not need to get too complicated. Let us go back to the verbose way. Let us retain the use of swap. To reverse an array a, I would have written

```
# reversed 1(b) where n = len(a)
na = n // 2
i = 0
while i < na:
        a[i], a[n - i - 1] = a[n - i - 1], a[i]
        i = i + 1
```

Frankly, I have to admit that reversed 1(b) is an overkill; reversing an array is too simple a chore to justify a verbose algorithm that I would gladly go for reversed 1(a). You may be required to examine and to do many stuffs on each individual object as you reversing the list. In those circumstances, as we have discussed in previous sections, one should be prepared to be verbose to produce a cleaner code.

To reverse the tuples, I might have written:

```
# reversed 2(f) where n = len(a)
i = n - 1
while a[i] in reversed(a) and i >= 0:
        # perform operation on (i, a[i]) ...
        i = i - 1
```

I like this verbose version as much as reversed 2(c). This code [reversed 2(f)] has one major advantage: with this code, I have greater control over the things I want to do on the reversed tuples if the requirements move again; and I would have gotten myself ready to spring into action. If the managers find the verbose code stale, only then I will inject a lot of idioms into it.

LAMBDA

If you have a series of temperatures, in a list and in Celsius, and you want to convert each of them into Fahrenheit, and on top of that insisting on using *map* and *lambda*, then I have a solution for you:

```
# e.g., temperatures = [-30, -10, 0, 20, 70]  # in Celsius
map(lambda x: (float(9)/5)*x + 32, temperatures)
```

Similarly, if you have a series of weights in pound (*lb*) and you want to convert each of them into kg, insisting on using map and lambda, then I have a solution for you as well:

```
# e.g., weights_in_lb = [20, 30, 40, 45, 50]
map(lambda x: x*0.453592, weights_in_lb)
```

Both solutions use lambda correctly, that is they use lambda as a *nameless function*. As expected, the use lambda will be, in most circumstances, less expressive than the use of named function. Whether the use of lambda will be a plus depends on the circumstances. If you were working in a meteorological department, where almost everybody knew the temperature conversion formula, I would say that the first solution should be unambiguous. In the second solution, there is no conversion formula; instead, there is a conversion factor, a floating-point number. I have met people who memorize conversion factors to do their jobs effectively. It might be still okay if everybody knew the conversion factor when they were looking at it. However, code is usually written by coders who are neither meteorological nor dietary experts. In those circumstances, both solutions can be sources of bugs. For example, a coder is asked to convert the temperatures back to Celsius, he or she may use the same formula. Of course, both solutions can be made less ambiguous by commenting the code, but I am still not entirely happy about that.

Sadly, once you try to associate a name with a lambda function, you are in effect defeating the purpose of using lambda in the first place. Let us use the previous example and rewrite it as follows.

```
lb_to_kg = lambda x: x*0.453592
map(lb_to_kg(x), weights_in_lb)
```

This code is much clearer than the previous one; it states clearly what the lambda function does, as the function name suggests, that is to convert *lb* to kg. Unfortunately, the code **does not work**, because map works only with lambda. If you apply enough pressure to force me to use lambda, I might inadvertently solve the problem this way:

```
lb_to_kg = lambda x: x*0.453592
map(lambda x: lb_to_kg(x), weights_in_lb)
```

I have used the keyword lambda twice! We should hate repetition of this sort. Let me get rid of one lambda:

```
def convert_lb_to_kg(x):
    return x*0.453592

    map(lambda x: convert_lb_to_kg(x), weights_in_lb)
```

This is a better-looking solution.

The two definitions, `lb_to_kg` and `convert_lb_to_kg`, are the "same" function. If you use them without looking inside, you will not be able to tell the difference. Let us instruct two Pythonists who are in love with list comprehension to write the conversion, using the conversion functions. You will get

Using `lb_to_kg` function defined earlier
```
[lb_to_kg(x) for x in weights_in_lb]
```

Using `convert_lb_to_kg` function defined earlier
```
[convert_lb_to_kg(x) for x in weights_in_lb]
```

Both of these solutions are quite clear, the first using a "named" lambda and the second using the function definition. It is very difficult for me to argue my case, once a function is put into use and treated as black box!

Fortunately, the following code **does not work**:

```
[lambda x: x*0.453592 for x in weights_in_lb]
```

If this code had worked, I would be upset. Imagine reading this line of code without a comment from the code writer for the first time. The code starts with `lambda`, a meaningless word, which is followed by a string of meaningless characters until it reaches `for`. The lambda, without doubt, impedes comprehension.

The use of lambda in Python can be confusing if we do not stick tightly to the definition of lambda as a nameless function. Let say you want to decide which of the two numbers, x or y, is larger:

```
the_larger_number_is = lambda x, y: x if x > y else y
```

The code is quite clear, even though the use of lambda troubles me a bit, because it is associated with a name. If you compute the answer only once, why not use a conditional expression instead,

```
the_larger_number_is = x if x > y else y
```

The use of lambda in recursion should trouble you even more. A typical implementation of the recursive Euclid algorithm is given below:

```
def gcd(a, b):
    if b == 0:
        return a
    return gcd(b, a % b)
```

Compared it to the following lambda version:

```
gcd = lambda a, b: a if b == 0 else gcd(b, a % b)
```

I am worried about those Python programmers who fell in love with this kind of abstraction. It will be a sad day for Python if lambda becomes idiomatic this way.

Allow me to reinforce my point with another example. Let us use lambda to rebuild a primality test function for small numbers (not a good algorithm, no matter, because I use it for illustration). First, the recursive version:

```
def is_prime(n, d =3):
```

```
    if n % 2 and n > 3 and n % d:
        if d*d > n:
            return True
        else:
            return is_prime(n, d = d + 2)
    else:
        return n == 2 or n == 3
```

Now, let us rewrite it using lambda:

```
is_prime = lambda n, d = 3: (True if d*d > n else
is_prime(n, d = d + 2)) if n % 2 and n > 3 and n % d else
n == 2 or n == 3
```

This primality test function is the simplest among the primality tests I know. Imagine writing functions using lambda that are more complicated than this! I would argue that the only reason you do it is to confuse your fellow coders. The great van Rossum expressed his view on lambda in his blog (van Rossum, 2005). With the popularity of functional programming on the rise lately, he might have changed his mind. Functional programming requires lambda as its essential tool. This chapter is not an avenue to explore this issue; I will write about this in the future.

SLICING

Slicing is on the fast-track route to become Pythonic. I do not know the exact number, but whenever I have the chance to read Python code, more and more slicing seems to appear. Personally, I do not mind slicing because it does not bug me that much compared to other idioms featured in this chapter, but I do wish Python authors could explain slicing much better when they write about it. In this chapter, we apply slicing to obtain a substring from a string.

Python has an odd way to index an array of things (so are some other languages). One can count from the beginning, that is counting forward, and start with an index 0; or one can count from the end, that is counting backward, and start with an index -1. Let us examine the string

```
s = "PituQmEcKLj"
```

The two indexing systems at play for the string is shown in the following figure.

0	1	2	3	4	5	6	7	8	9	10
P	i	t	u	Q	m	E	c	K	L	j
-11	-10	-9	-8	-7	-6	-5	-4	-3	-2	-1

Figure 3. Two index systems for "PituQmEcKLj".

Consequently, for $i = 0, 1, 2, \ldots, n-1$, where n is `len(s)`, we have

$$s[i] = s[i - n].$$

Each character of the example has two indices. For example, 'K' is indexed by either 8 or -3. So, the following snippet will print True eleven times.

```
# n = len(s)
for i in range(n):
    print(s[i] == s[i-n])
```

Python programmers use slicing to obtain specific subarray of an array. Using the string s as an example, the syntax of slicing is

$$s[\ begin : end : step\ ]$$

When you read slicing code, you have to be aware of the following issues:

- The slice itself. This is elementary. The substring begins with the character indexed by *begin*, but ends with character **before** the one indexed by *end*. So, `print(s[2:6:1])` will print `tuQm`. The *step* specifies steps taken when you jump from begin to end.
- Default values of the indices if you leave them out. You can leave out *begin*, *end*, or *step*. If you leave any of them out, the splicing assumes **default value** for that argument. The following snippet prints the string `PituQmEcKLj` nine times.

```
print(s)              # print the string
print(s[:])
print(s[::])
print(s[0:])
print(s[0:n])         # n = len(s)
print(s[:n])
print(s[-n:n])
print(s[-n:])
print(s[0:n:1])
```

However, `print(s[])` will be a syntax error (I hope you know why). Once you use slicing, you must at least use one colon. The default value for *step* is 1. The default value for *begin* is 0 or -11, depending on the *step*. The default value for *end* is 11, or `len(s)`. You have to be careful that the default *end* is not -1, or -0! Therefore, the print statement `print(s[:-1])` will print `PituQmEcKL` without the '`j`'.
- The complication of mixing the two indexing systems. You can mix the two indexing systems, but complication assures. The following snippet prints `iumc` four times.

```
print(s[1:8:2])
print(s[1:-3:2])
print(s[-10:8:2])
print(s[-10:-3:2])
```

The *step* has life of its own, which I will explore next.

- The meaning of negative *step*. This is another elementary one. When you specify the *step* to be a negative value, it means that the substrings you get will reverse part of the original string. The following snippet prints `LEu` four times:

```
print(s[9:1:-3])
print(s[9:-10:-3])
print(s[-2:1:-3])
print(s[-2:-10:-3])
```

- The influence of *step* on other indices. The fifth issue is when you specify the value of *step*, its sign affects the default values of *end* but not *begin*. The following snippet prints the string in reverse.

```
print(s[::-1])
print(s[-1::-1])
print(s[10::-1])
print(s[10:-12:-1]) # end = -12
```

As you might have guessed, the default for *end* is -12, or -(len(n) + 1), once you have decided to traverse backward. Consider the following snippet, which prints the string `PituQmEcKLj` as usual.

```
print(s[::1])
print(s[0::1])
print(s[-11::])
print(s[:11:]) # end = 11
```

As you might have guessed, the default for *end* is now `11` or `len(n)`, once you have decided to traverse forward.
- The decision. You can either use forward, backward, or mix index systems. In many circumstances, mixing of the two indexing systems will reduce the readability of your code, but, as usual, there are always exceptions. I challenge you to think of an example

where mixing of these indexing systems is better than not mixing them.

EXCEPTIONS AND EAFP

The phrase "never catch all exceptions" is not unique to Python. Programmers from other languages knew this praxis as well; see, for example, (Haggar, 2000; Bloch, 2001; Bloch, 2008; Bloch, 2018). Advice given on the use of exception handling by Bloch, a java guru, over three editions of his books (Bloch, 2001; Bloch, 2008; Bloch, 2018) are quite immutable; so, it is not too expensive an investment to learn how to use exception effectively, even though the advice is not in Python. I dare not say exception catching is language neutral and universal, but I dare say good coding practices are universal.

Pythonists are generally leading EAFP lives. Nevertheless, EAFP may lead to ignorance. Let us use the string `s = "PituQmEcKLj"` again and write an example that displays a random character from the string based on an input integer, which will be the index to obtain a character from the string. We know that the string has two indexing systems. Therefore, if `n = len(s)`, any integer that satisfies the following condition will be a valid index for the string.

```
random.randint(0, 2*n - 1) - n
```

Let us create a function `random_character()` that keeps on issuing a random character from an input integer, call it `random_num`, which is randomly fed from another system, until you switch it off . The line of code is something like this:

```
random_character(s[random_num])
```

An `IndexError` exception, which many smart people ask you to ignore, will be thrown in no time if you let EAFP take its course. When the exception is raised, your code is in effect executing a jump, similar to `goto`. If you want the system to continue issuing a random character from `s`, you have to restart the function again. An additional `if`-statement, which uses an idiomatic sugar, would have prevented this exception from being thrown:

```
if -n <= random_num <= n - 1:
   random_character(s[ramdom_num])
```

In short, do not let EAFP take over your thinking. Solving a problem means understanding the problem. EAFP can be dogmatic. In many situations, you may let EAFP take its course, but not in all situations. Without using a single try-catch anywhere in this section, exception is handled or rather prevented.

WHAT I WOULD LIKE TO COVER BUT DIDN'T

So many. The use of *any*, *all*, *in*, *dict*, *zip*, *extend*, *set*, etc. Endless. By now, you should have gotten used to the way I argue my case. If I continue writing about the functions that I italicized, I will be repeating the same patterns of argument. Yet, before leaving this chapter, let me highlight one of the interesting entities of Python.

Python snippets I presented in preceding sections contain list comprehensions that immediately compute but do not store or build objects. Therefore, I must cover generator expressions (GE), which are featured in PEP 289. Many of us cannot distinguish between GE and list comprehension, except the use of parentheses in GE and square brackets in list comprehension. And this imperfect knowledge creates a veneer that hides our weakness and may lead to potential buggy creations.

Let us write a snippet of an allotment application. The application is supposed to select randomly a fixed number of objects from a list. The application passes back a generator object to another application to be processed, and at the same time removes all the selected elements from the original list. The choice of generator object is superb, because this object, unlike list comprehensions, does not compute straightaway but build like a persistent object. To create the generator object with 10 elements from the list a is an easy one-liner, even though I do not like the use of variable x and the repeating use of a:

```
g = (a.pop(a.index(random.SystemRandom().choice(a)))
    for x in range(10))
```

Very Pythonic indeed, but there is one problem: the list a is passed by value and not by reference. The original list has not been modified! Since the receiving function *insists* on using a generator object, and refuses to budge, and to get things done swiftly, I write

```
b = []
for i in range(10):
    rand_choice = random.SystemRandom()
    temp = rand_choice.choice(a)
    b.append(temp)
    a.remove(temp)
g = (b)
```

This code is no longer a one-liner, and its cool factor has gone down many scales compared to the previous one. What really matters is in the end I created the required generator object and modified the original list. Spending another 10 minutes, I refactored the code and created a function with two parameters, which are the original list (passed by reference, to be sure) and the number of elements selected from the original list. If I had the time and I found out that my fellow coders were studying the code, I might transform it to beautiful, idiomatic Python to prevent my ego from being punctured. However, if the function is being used without complaint, it is likely that I will not bother to do any modification. Believe it or not, you can write a whole book about GE.

Where does the drive to write the "beautiful" code in idiomatic Python come from? The answer is rather speculative, but I will try anyway. It is likely that the urge to write idiomatic Python comes from emotion. This is because as a sports enthusiast, I see it displayed undoubtedly in both winning and losing teams. One of the emotions, pride, which is a subcategory of joy according to Fischer et al. (Fischer, Shaver, and Carnochan, 1990), is the best candidate, because it is the self-conscious emotion that we develop when we deal with the stricture of the environment (Scheff, 1990), and in our case the Pythonic environment. Pride has its root in chemistry (Vikan, 2017). In a recent book I read about the company culture that brings the best out of people by Sinek (Sinek, 2017), two of the five neurochemicals listed that caught my attention are dopamine and serotonin. These two chemicals perhaps are the culprits that bliss us out when we successfully write the code in "beautiful" idiomatic Python. I hardly can do justice here for the study of neurochemistry in idiomatic Python.

In professional software development, if you ever get a chance to work for one, you will see plain code you see in this chapter, textbooks, StackOverflow, etc. Imagine that you have thousands of developers working for the same project, one of the many attempted ways to produce acceptable buggy code is to have a comprehensive software testing regiment in place. Maguire (Maguire, 1993) has documented his experience in writing code during his tenure with Microsoft. Programming Pearls (Bentley, 1999; Bentley, 1988) by Jon Bentley are evergreen – so read them. Martin's clean code (Martin, 2008) has been widely read, and so you should check it out. Python programming books inundate the bookstores, printed or otherwise. In addition to the Python books that I referred to earlier in this chapter, my favorite Python books are (Gorelick, 2014; Downey. 2015; Guttag, 2016; Overland, 2017). And I believe my favorite list will continue to change as new publications hit the bookstores.

The popularity of Python is rising in recent years. Writing code in Python means freedom: you will find many cool ways to do the same thing. It truly amazes me to read code produced by some innovative Pythonists, despite my objection to their pride. Whenever I write Python code, I remind myself that idiomatic Python can be the condiment of the code, but it should never be the only ingredient. I cannot stop thinking about the loops and the idiomatic Python in this chapter; they are but despicable entities to functional programmers.

APPENDIX A

Abbreviations used in this chapter:

- **BDFL** for Benevolent Dictator For Life
- **PEP** for Python Enhancement Proposal
- **EAFP** for It is Easier to Ask for Forgiveness than Permission
- **GE** for Generator Expressions
- **JSON** JavaScript Object Notation
- **ORM** for Object Relational Mapper

APPENDIX B

The following Python script was used to perform the experiment that produced the bar charts with error bars in Figure 1. An attempt has been made to reduce function call overhead, and therefore repetition of code is apparent in selection sort 1 and 2. Furthermore, I avoided NumPy and Panda. The plot was produced using the demo script from matplotlib (Matplotlib.org, 2018). Appendix C, despite coming from a different experiment, should give you a clearer picture of what would be the output of this script if you run it in your machine.

```
from random import random
import pathlib
import time
import math

def sattalo(file_name, the_size):
    p = pathlib.Path(file_name)
    p.touch()
```

```python
        if p.exists():
            p.unlink()
        arr = list(range(the_size))
        n = len(arr)
        i = n
        while(i > 1):
            r = int((i - 1)*random())
            temp = arr[r]
            arr[r] = arr[i - 1]
            arr[i - 1] = temp
            i = i - 1
        file = open(file_name, "w")
        for i in range(n):
            file.write(str(arr[i]) + '\n')
        file.close()

def read_data(file_name):
    with open(file_name) as file:
        a = file.read().splitlines()
    file.close()
    a = list(map(int, a))
    return a

def selection_sort1(the_size, data_file, result_file, n_repeat):
    rst = list(range(n_repeat))
    f = open(result_file, 'a')
    f.write("\n== Idiomatic Swap == \n Size = " + str(the_size))
    for k in range(n_repeat):
        a = read_data(data_file)
        n = len(a)
        start = time.time()
        for i in range(0, n - 1):
            p = i
            for j in range(i + 1, n):
                if(a[j] < a[p]):
                    p = j
            if(p != i):
                a[p], a[i] = a[i], a[p]
```

```python
        end = time.time()
        f.write('\n' + str(end - start))
        rst[k] = end - start
    mean = sum(rst)/n_repeat
    s = 0
    for i in range(n_repeat):
        s = s + (rst[i] - mean)**2
    f.write("\n Mean = " + str(mean))
    f.write("\n std = " + str(math.sqrt(s/(n_repeat - 1))))
    f.close()


def  selection_sort2(the_size,  data_file,  result_file,
n_repeat):
    rst = list(range(n_repeat))
    f = open(result_file, 'a')
    f.write("\n== temp variable Swap =\n Size = " +
str(the_size))
    for k in range(n_repeat):
        a = read_data(data_file)
        n = len(a)
        start = time.time()
        for i in range(0, n - 1):
            p = i
            for j in range(i + 1, n):
                if(a[j] < a[p]):
                    p = j
            if(p != i):
                swap = a[p]
                a[p] = a[i]
                a[i] = swap
        end = time.time()
        f.write('\n' + str(end - start))
        rst[k] = end - start
    mean = sum(rst)/n_repeat
    s = 0
    for i in range(n_repeat):
        s = s + (rst[i] - mean)**2
    f.write("\n Mean = " + str(mean))
    f.write("\n std = " + str(math.sqrt(s/(n_repeat - 1))))
    f.close()
```

```
sort_sizes =
    [5000,6000,7000,8000,9000,10000,20000,30000,   40000,
50000]
data_file_name = "Data.txt"
result_file_name = "Results.txt"
n_repeat = 20
f = open(result_file_name, 'a')
num = 0
for sort_size in sort_sizes:
    sattalo(data_file_name, sort_size)
    num = num + 1
    if num % 2:
        print("Selection 1 followed by 2")
        selection_sort1(sort_size,
            data_file_name,result_file_name,n_repeat)
        selection_sort2(sort_size,
            data_file_name,result_file_name,n_repeat)
    else:
        print("Selection 2 followed by 1")
        selection_sort2(sort_size,
            data_file_name,result_file_name,n_repeat)
        selection_sort1(sort_size,
            data_file_name,result_file_name,n_repeat)
```

APPENDIX C

The Python script to obtain the physical running time for finding the minimum of an array is similar to that of the script presented in Appendix B. An average Python programmer should know how to change the script in Appendix B to obtain the running time. Be reminded that this result was obtained using my machine.

Table 2. Results of running time performed on my machine to find the minimum of an array of random numbers using two different ways: the built-in min function and the verbose way using Python code. Only two sizes of the arrays are given here: 100,000 and 1,000,000

Using min(a)	Using Verbose	Using min(a)	Using Verbose
Size = **100,000**	Size = **100,000**	Size = **1,000,000**	Size = **1,000,000**
0.0	0.0	0.015652418136	0.046913862228
0.001028060	0.0	0.015652179718	0.046912908554
0.015647411	0.0	0.015628337860	0.046916007995
0.0	0.0	0.015627861022	0.046880483627
0.00203180	0.0	0.015626430511	0.046912908554
0.0	0.0	0.015658378601	0.032771110534
0.0	0.0050377	0.015656471252	0.046901226043
0.01562094	0.0	0.015626668930	0.046909093856
0.0	0.0	0.015654563903	0.062534093856
0.0	0.0	0.015653610229	0.046910285949
0.0	0.0	0.015656471252	0.046914815902
0.0	0.0	0.015659570693	0.046915769577
0.0	0.0	0.015624761581	0.046916723251
0.0	0.0	0.015659809112	0.057926893234
0.0	0.0	0.015656709671	0.046908617019
0.0	0.0	0.0	0.053681373596
0.0	0.0	0.015653610229	0.045646667480
0.0	0.0040404	0.015626192092	0.038550615310
0.0	0.0156538	0.015628814697	0.046914100646
0.0	0.0	0.015627145767	0.046909570693
Mean = 0.00171	Mean = 0.00123	Mean = 0.01486	Mean = 0.04739
std = 0.004785	std = 0.0036718	std = 0.0034980	std = 0.0059739

ACKNOWLEDGMENTS

To the anonymous reviewer who read the chapter and gave valuable suggestions to improve it, I really appreciate your input and effort. Thanks are due to Mr. K. K. Tey for spending time on reading and discussing some of the idiomatic Python code of this chapter. The support from Dr. R. P. Manas, Dr. W. O. Choo, S. M. Ng, S. F. Lim, and Y. M. Tan is invaluable; these people rock. Thanks to my family for leaving me alone to read books. The Python community deserves my thanks as well. Pythonists are generally kind and humble.

REFERENCES

Aho, A. V., Hopcroft, J. E. & Ullman, J. D. (1983). *Data Structures and Algorithms*, Reading: Addison-Wesley.

Bentley, J. (1999). *Programming Pearls*, Second Edition. Crawfordsville: Addison-Wesley Professional.

Bentley, J. (1988). *More Programming Pearls: Confessions of a Coder*. Crawfordsville: Addison-Wesley Professional.

Blanks, H. (2011). *PEP 20 (The Zen of Python) by Example* [WWW Document]. URL http://artifex.org/~hblanks/talks/2011/ pep20_by_example.html (accessed 5.24.2018).

Bloch, J. (2001). *Effective Java™: Programming Language Guide*. Upper Saddle River: Addison-Wesley.

Bloch, J. (2008). *Effective Java™*, Second Edition. Upper Saddle River: Addison-Wesley.

Bloch, J. (2018). *Effective Java™*, Third Edition. Upper Saddle River: Addison-Wesley.

Cython.org. (2018). *Cython C-Extensions for Python* [WWW Document]. URL http://cython.org/ (accessed 5.24.2018).

Dobelli, R. (2013). *The Art of Thinking Clearly*. London: Sceptre, Chapter 92.

Downey, A. B. (2015). *Think Python: How to Think Like a Computer Scientist*, Second Edition. Sebastopol: O'Reilly Media.

Fischer, K. W., Shaver, P. R. & Carnochan, P. (1990). *How Emotions Develop and How they Organise Development*. Cognition & Emotion 4, 81-127. https://doi.org/10.1080/02699939008407142.

Gladwell, M. (2007). *Blink: The Power of Thinking Without Thinking*. Hachette UK.

Gladwell, M. (2009). *Outliers*. London: Penguin UK.

Goodger, D. (2008). *Code Like A Pythonista: Idiomatic Python* [WWW Document]. URL http://python.net/~goodger/projects/pycon/ 2007/idiomatic/handout.html (accessed 3.2.2018).

Gorelick, M. & Ozsvald, I. (2014). *High Performance Python*. Sebastopol: O'Reilly Media.

Guttag, J. V. (2016). *Introduction to Computation and Programming Using Python: With Application to Understanding Data*, Second Edition. Cambridge: MIT Press.

Haggar, P. (2000). *Practical Java™ Programming Language Guide*. Upper Saddle River: Addison Wesley Longman.

Hettinger, R. (2013). *Transforming Code into Beautiful, Idiomatic Python* [WWW Document]. YouTube. URL https://www.youtube.com/watch?v=OSGv2VnC0go (accessed 3.2.2018).

Hettinger, R. (2013). *Beyond PEP 8 - Best Practices for Beautiful Intelligible Code* [WWW Document]. YouTube. URL https://www.youtube.com/watch?v=wf-BqAjZb8M.

Jaworski, M. & Ziadé, T. (2016). *Expert Python Programming*, Second Edition. Birmingham: Packt Publishing.

Kaleniuk, O. (2017). *Going beyond the idiomatic Python* [WWW Document]. URL https://hackernoon.com/going-beyond-the-idiomatic-python-a321b6c6a5e6 (accessed 3.2.2018).

Kaufman, J. (2014). *The First 20 Hours*. Portfolio Trade.

Knupp, J. (2013). *Writing Idiomatic Python 3.3*. CreateSpace Independent Publishing Platform.

Krogerus, M. & Tschäppeler, R. (2018). *The Communication Book*. Portfolio.

Liguori, R. & Liguori, P. (2014). *Java Pocket Guide: Instant Help for Java Programmers*, Third Edition. Sebastopol: O'Reilly Media, p. 4.

Maguire, S. A. (1993). *Writing Solid Code*. Redmond: Microsoft Press.

Martin, R. C. (2008). *A Handbook of Agile Software Craftsmanship*. Upper Saddle River: Prentice-Hall.

Matplotlib.org. (2018). *api example code: barchart_demo.py* [WWW Document]. URL https://matplotlib.org/examples/api/ barchart_demo.html (accessed 5.24.2018).

Nanjekye, J. (2018). *Python 2 and 3 Compatibility*. New York: Apress.

Norvig, P. (2014). *Teach Yourself Programming in Ten Years* [WWW Document]. URL http://norvig.com/21-days.html (accessed 3.2.2018).

Oldham, J. D. (2005). *What happens after Python in CS1? Journal of Computing Sciences in Colleges*, 20(6), 7-13.

Orfanakis, V. & Papadakis, S. (2016). *Teaching basic programming concepts to novice programmers in Secondary Education using Twitter, Python, Ardruino and a Coffee Machine*, in: Proceedings of the Hellenic Conference on Innovating STEM Education (HISTEM), University of Athens, Greece. 2016. Available at: http://stemeducation.upatras.gr/ histem2016/ assets/ files/ histem2016_ submissions/ histem 2016_paper_8.pdf (accessed 5.24.2018).

Overland, B. (2017). *Python without Fear*. Upper Saddle River: Addison-Wesley Professional.

Peters, T. (2004). *PEP 20 - The Zen of Python*, [WWW Document]. URL https://www.python.org/dev/peps/pep-0020/ (accessed 5.24.2018).

Ramalho, L. (2015). *Fluent Python*. Sebastopol: O'Reilly Media.

Reitz, K. & Schlusser, T. (2016). *The Hitchhiker's Guide to Python: Best Practices for Development*. Sebastopol: O'Reilly Media. Chapter 4, pp. 43 – 60.

Sattolo, S. (1986). *An algorithm to generate a random cyclic permutation*. Information Processing Letters 22, 315-317. https://doi.org/10.1016/0020-0190(86)90073-6.

Scheff, T. (1990). *Socialisation of emotions: Pride and shame as causal agents*. In: Kemper T. D. ed. Research Agendas in the Sociology of Emotions, Albany, New York: State University of New York Press, pp. 281–304.

Sinek, S. (2017). *Leaders eat last: Why some teams pull together and others don't*. New York: Portfolio/Penguin.

Skrien, D. (2009). *Object-Oriented Design Using Java*. New York: McGraw-Hill Higher Education.

Slatkin, B. (2015). *Effective Python*. Indianapolis: Pearson Education.

SQLAlchemy. (2018). *SQLAlchemy - The Database Toolkit For Python* [WWW Document], n.d. [WWW Document]. URL https://www.sqlalchemy.org/ (accessed 6.29.2018).

Tracy, J. (2016). *Pride: The Secret of Success*. Boston: Mariner Books, Houghton Mifflin Harcourt.

van Rossum, G., Warsaw, B. & Coghlan, N. (2001). *PEP 8 -- Style Guide For Python Code* [WWW Document]. URL https://www.python.org/dev/peps/pep-0008/ (accessed 6.28.2018).

van Rossum, G. (2005). *The fate of reduce() in Python 3000, In: All Things Pythonic*.[WWW Document]. URL http://www.artima.com/weblogs/viewpost.jsp?thread=98196 (accessed 3.2.2018).

Vikan, A. (2017). *A Fast Road to the Study of Emotions: An Introduction.* Switzerland: Springer International Publishing, Chapter 12.

Warsaw, B. (2010). *Import this and the Zen of Python* [WWW Document]. URL https://www.wefearchange.org/2010/06/import-this-and-zen-of-python.html (accessed 5.24.2018).

Weinberg, G. (1971). *The Psychology of Computer Programming*. New York: Van Nostrand Reinhold Company.

Date Submitted: March 5, 2018. Date Accepted: July 6, 2018.

In: Current STEM. Volume 2　　　　ISBN: 978-1-53616-042-0
Editor: Maurice HT Ling　　　　© 2019 Nova Science Publishers, Inc.

Chapter 2

A REFLECTION OF MY 4 YEARS OF UNDERGRADUATE STUDY IN SINGAPORE: WHAT I LEARNT FROM SINGAPORE

*Argho Maitra**

Department of Applied Sciences, Northumbria University,
Newcastle, UK
School of Life Sciences, Management Development Institute
of Singapore, Singapore

ABSTRACT

In this narrative, I shall chronologically account for all of my experiences in Singapore, stretching through a span of four years. As would appear on surface, the narration will provide the reader a glimpse of the following events in my life: my background in India, my entry into Singapore, the onset of diploma to the eventual attainment of an honours degree. However, the ultimate objective of this narrative is to explicitly make the reader aware of the profound contribution that Singapore has made to my life. I can positively reflect from my experiences that (1) the acquisition of knowledge *per se* is of trivial value if it is not applied in a research setting, (2) the indispensable guidance provided by my mentor

---

* Corresponding Author's E-mail: argho.official@gmail.com.

went beyond the scope of the research project and was conducive to bolster my passion for science, (3) an implementation of a research-based academic curriculum is of paramount importance relative to a text-book based education and (4) embracing my tryst with adversity allowed me to better perform throughout the course of my undergraduate studies. Studying in Singapore not only engrained in me the multifaceted aspect of research but also made me appreciate life more. As a consequence, I became steadfast in my pursuit to become a scientist.

BACKGROUND

My great-grandfather, Mr N G Banerjee, was a scientist-analytical chemist at the Central Institute of Mining and Fuel Research (CFRI) in Dhanbad, India. He was sent to Poland as part of the Indian scientists delegation, where he mastered written Polish and was appointed the Polish paper translator for the Indian National Scientific Documentation Centre (INSDOC). My grandfather, Mr D. K. Bhattacharyya, too, was a scientist-geologist at CFRI and my grandmother, Sunanda Bhattacharyya, finished her graduate studies in philosophy-cum-world history. My mother finished her Master's in Organic Chemistry and dedicated her life to full-time teaching. Hence unbeknownst to me, since childhood, I had a predisposition towards an enquired mind and dealt somewhat calmly with educational stress and anxiety. Several studies examining the relationship between maternal parental education and educational outcome of a child have indicated a linear correlation between both the variables along with an increased likelihood in the willingness of the child to succeed (Pufall et al., 2016, Carneiro et al., 2013, Augustine, 2017, Dickson et al., 2016). Similar research focused on the disparity between first-generation (first in family to attend higher studies) and continuing-generation (at least one parent completed higher studies) academics stated that the former not only had a poor chance of dealing with educational stress and anxiety but also showed deteriorated academic performance, relative to the latter (Janke et al., 2017, Harackiewicz et al., 2014, Jury et al., 2015, Adams et al., 2016, Tibbetts et al., 2018). Thus, my parents' penchant for education motivated me to pursue my secondary school studies at one of the most prestigious boarding schools in India - Birla Vidya Mandir (BVM), situated atop the picturesque hills of Nainital. It was a turning point in my life as BVM not only broadened my horizons but also made me resilient. BVM's motto, transliterated from Sanskrit, inculcates *"svadharme nidhanaṁ śreyaḥ"* (3:35) (*The Bhagavad-gita*, 1929), which derives from the Bhagavad Gita (a Sanskrit epic and a storehouse of spiritual wisdom) and is interpreted as follows: *"it is better to follow one's own principles, however faulty, than follow an alien law well-wrought out"* (Kalra et al., 2018, Kalra et al., 2017). Of all the things that BVM taught me, individualism was at the core of it; BVM taught me to embrace my idiosyncrasies. From scratch, I 'relearnt' the essence of life, something which I had learnt years ago, albeit imperfectly. I was a dominant presence in debate, dramatics, quizzes, elocution, sports, singing, dance, just to name a few. I was neither the 'Jack' in James Powell's *"All work and no play makes Jack a dull boy"* nor the one in Maria Edgeworth's *"All play and no work makes Jack a mere toy"* (Edgeworth, 1827, Howell, 1659). I had rather learned to be an amalgam of both. By the time I had finished my secondary school education in 2008, I had culminated an interest in pure sciences, linguistics and information technology. Consequently, I received an aggregate of 94.6% in the CBSE X[th] board examinations (equivalent to GCE O-levels) and ranked 5th overall in my school with considerably high marks in mathematics (90), science (94), information technology (100) and Sanskrit (93).

Unfortunately, I had to cease my educational endeavour at BVM, where I had dearly sought to finish my senior secondary education. My father had an accident: he fractured his right arm and developed a skin and soft tissue infection on his right lower limb inasmuch that he was left immobile and could not perform his day-to-day activities, and which also lead to the termination of his employment. In addition, my mother suffered from type 2 diabetes and hypertension. Devoid of a silver spoon in my mouth and worried about the well-being of my family, I became hard-pressed to pursue my senior secondary studies back home. I took care of my father as we struggled both socioeconomically and psychologically. By the time I had completed schooling in 2010, there had been a significant decline in my performance from the CBSE Xth to the CBSE XIIth board examinations (equivalent to GCE A-levels), though I had managed to secure a first class. Soon my father's condition worsened as he developed severe infectious cellulitis in his right lower limb, which lead to the complete cessation of his working capacity. Generously though, the director of my father's workplace, Mr Ahad Ullah, Globe Leather Industries, understood my circumstances and gave me an opportunity to intern in my father's stead in the marketing sector. It was overwhelming at first as I had no experience whatsoever in sales and marketing, however, my time-efficient solutions to critical problems, analytical thinking, meeting of deadlines under stringent time constraints and presentation skills earned me a permanent job as a 'Marketing Executive' lasting four years (2010-2014). Following is an excerpt from Mr Ullah's referee letter in which he recommended me to pursue higher studies –

> "I not only found him hard-working and responsible but he also has a great sense of business. He is extremely cooperative and possesses the capacity to contribute positively to his area of leadership. He is ambitious in life and wants to achieve something special. His extraordinary ability to analyse problems and outline the necessary courses of action is invaluable."

Throughout my marketing tenure, I had competed in several amateur bodybuilding shows and in one of which I bagged the runner-up prize for the 'Mr Kanpur Bodybuilding Contest' in Kanpur. Leading up to this exact point in time, I had already developed a predilection for anatomy, physiology, and nutritional sciences; I was fascinated by how the human musculature undergoes atrophy and hypertrophy in the absence and presence of a balanced nutrition. Although my bodybuilding career was ephemeral, the subtle nuances of human anatomy and physiology sparked an interest in me. Inspired by my grandfather and great-grandfather, and toughened by all the adversities that I had faced, I never quite lost sight of what I had initially set out to achieve, i.e., knowledge *per se*. Several studies on the relationship between adversity quotient and academic achievement concluded that the degree of adversity might not have a direct influence on academic achievement, however, it does have a positive association with increased persistence and emotional intelligence, which further leads to better chances of success (Mohd Matore et al., 2015, Verma et al., 2017, Zamri Khairani and Syed Abdullah, 2018, Miri et al., 2013, Ranasinghe et al., 2017). In accordance with (Woodward et al., 2017), my eagerness for biology derived from an interplay of different reasons. Growing up under the shadow of my grandfather, also a diabetic, I always had an instinct to educate myself about nature and science, and his subsequent demise brought about by myocardial infarction developed an urgency in me to learn about the condition and what caused it. The death of my grandfather and the illness' of my parents were therefore decisive factors that strengthened my outlook on biology as well as the medical sciences. Thus, my profound passion for biology prompted me to study medical sciences. Since I had not taken biology in senior secondary, I was ineligible to pursue medicine or any other biological discipline in India. In the autumn of 2014, my maternal uncle who had come by for a visit, handed me over the prospectus of the Management Development Institute of Singapore (MDIS) and discussed the possibility of my undergraduate studies in Singapore. He wanted me to pursue business and marketing in relation to my job experience, however, going over the curriculum for both the biomedical and biotechnological sciences at MDIS, I became fixated on studying science. So in 2015, through fortuity, I got a unique opportunity to pursue my undergraduate studies in Singapore. My interest in medical sciences demanded high school biology as a prerequisite, however, the lack of biology amongst my core courses in senior secondary would eventually compel me to do a Foundational Diploma in Biomedical Sciences at MDIS in Singapore. Although my parents never set any expectations, three days prior to my departure, my father made me realise that I have to become my family's backbone and in order for that to happen, I must work hard despite the hurdles that I will face. With that imprinted in my mind, I left for Singapore.

THE FIRST THREE YEARS IN SINGAPORE

I believe my entry into Singapore to be both serendipitous and a 'calling'; Singaporean education not only amplified my passion for biological sciences but also life skills. Singapore has a fast-paced lifestyle; it is habitual to see people reading novels, studying or attending to their emails while holding a cup of coffee and simultaneously walking on the pedestrian walkway. This was intimidating at first as the culture where I come from is quite the opposite, however, I embraced this change as it made my settling down relatively easier. As I reflect on my experiences, I recall a scene from *Back to the Future Part III* (1990), where Emmet 'Doc' Brown reminds Marty McFly *"You're just not thinking fourth dimensionally,"* since Marty fails to see the bigger picture (Zemeckis, 1990). Now in retrospect, it will suffice to say that Singapore pushed me out of my comfort zone and reprogrammed me to think fourth dimensionally. Significantly less than one-fourth the size of India, Singapore, since its independence has established itself as the 'biomedical research powerhouse' (Van Epps, 2006, Samarasekera et al., 2015, Chuan, 2007, Nature, 2010). Flipped classroom learning and gamification are two of the many methods forming the basic framework of the innovative student learning process (Lum et al., 2018, Samarasekera et al., 2015) in the Singaporean education system. This forced me to articulate beyond the pedagogical humdrum of the normal classroom. As compared to my counterparts who were nearly 6-7 years younger, I entered my diploma at a significantly older age of 24, nonetheless, I received a very warm reception and made friends from different parts of the world. Mathematics, Physics, Chemistry, Anatomy, Physiology, Molecular Cell Biology and Genetics were the core modules in my diploma, whose holistic student assessment process equipped me with better professional skills. Of all the biomedical skills that I gained over the course of my six month diploma, critical analysis of medical literature and hands-on-laboratory work quickly became my favourites. After the successful completion of my diploma, I was elated to join Northumbria University's **undergraduate program** - Bachelor of Science (Honours) in Biomedical Sciences, which was being offered at the MDIS campus in Singapore. My first year modules were similar in construct to pre-medical courses – Applied Anatomy and Physiology, Biochemistry, Cell Biology and Genetics, just to name a few. They primarily constituted basic sciences and served as building blocks of what was to come later. Not only did I become adept at critical analysis of both primary and secondary literature but I also developed a sheer passion for scientific writing. By the end of the first year, I had become steadfast on venturing into the world of medicine. My conviction was strengthened to an even greater magnitude when I entered the second year and got exposed to some of the underlying clinical principles of biochemistry and metabolic disorders, biotechniques and practical molecular genetics, biology of disease, microbiology, and cellular and immunological methods. But it seemed that my tryst with hardship was not over yet; amidst my second year, I was informed that I might have to discontinue my studies and withdraw from Singapore due to financial difficulties. At that moment, I was absolutely shattered; it was as if all had gone to waste. Be that as it may, I had my fair share of experiences to know the difference between managing adversity and running away from it. I believe that management of adversity goes hand in hand with achieving the task at hand: there will always be situations that will obstruct one's progress however it is imperative for the individual to lean against the wind, resist it and keep moving forward (Higgins et al., 2012). My lecturer for Cellular and Immunological Methods, Dr Sukumar Ponnusamy, was of huge assistance as his personal anecdotes provided me an emotional stability. Hence, following in the footsteps of *'where there's a will there's a way,'* I kept my calm and composure, and gave my undivided attention to the task at hand.

    My level of complacency hitherto ranged from getting good grades in my undergraduate to becoming a successful doctor in the future, however all of this changed when I witnessed a talk by Dr Indira Nath. She is a prominent Indian immunologist responsible for making pioneering contributions in cellular immune responses and nerve damage in human leprosy (Nath et al., 2015). She came to Singapore as part of the group of 400 scientists attending the 'Commonwealth Science Conference 2017' symposium (Nath, 2017). In her talk at MDIS (Wong, 2017), she shared her insights on some of the new emerging technologies in medicine, and I reminisce vividly as I stood up during the Q&A sessions and asked her, *"What is the most important advice you can give to someone who wants to become a doctor?"* to which she replied without a hint of prejudice, *"There is need for more scientists, upon whose research and problem solving skills can the doctors capitalize."* It was at this moment that I had an epiphany. My ambition of becoming a doctor was largely misdirected. Generally, there are several underlying factors that motivate aspiring doctors to choose medicine, out of which the inherent inclination towards 'the desire to help others' is where all 'would-be' healthcare professionals converge; Hippocrates' ideology of 'the doctor as a healer' sets the common ground amongst applicants (Goel et al., 2018, Heikkilä et al., 2015, Woodward et al., 2017). I too have an 'interest in people,' but both research and simultaneous patient care as a focal point in the future seemed improbable to me, as I had already developed a strong liking for research through scientific writing. I had also given thought to consider being a physician scientist, however for similar reasons and reasons beyond the scope of this article, I decided to focus primarily on research. As outrageous as it may seem, much like the layman, initially I had thought of the doctor as a scientist. A doctor is a master of the 'known' whereas a scientist's bread and butter is to tinker meticulously with the 'unknown' (Freed, 2004, Smith, 2004) and that is what got me interested in the human body in the first place. As Smith (2004) put it, I wanted to be part of a group who *"brush their teeth on only one side of their mouth to see whether brushing your teeth has any benefit"*. Henceforth, I entered the final year of my degree with an uncompromising conviction on wanting to pursue research.

The 'Honours' Year

The final year of my degree was the stepping stone that ultimately made me visualise my ambition. One of the key modules in my first semester was the Scientific Literature Review, wherein I had to pedantically analyse primary literature, abstract pertinent data from a broader context, narrow it down contextually and encapsulate all the information therein to present it to a supervising committee. My passion in scientific writing facilitated me to thoroughly enjoy the process, whether it be the literature investigation or the completion of the review. Studies in positive psychology have indeed confirmed that the existence of a harmonious passion generates a stronger grasp and control over an activity and is positively correlated with an outpour of positive emotions and well-being (Briki, 2017, Vallerand, 2012, Verner-Filion et al., 2017, Bogg, 2016). Having developed a fascination for oncogenetics early on in the second year, 'cancer cell signalling' formed the fundamental core of my review, in which I critically analysed then current research on BCR-ABL mediated chronic myeloid leukaemia; the anomalous pathways disrupted by it and the therapeutic approaches necessary to eliminate its progression. The obvious motive of the literature review was to get the students ready for the psychologically demanding 'Research Project,' about to come in the second semester.

Before the start of the second semester, I recall opening an email containing a list of potential projects with the names of their respective supervisors, and what caught my attention was a section titled 'Genomics'. Genomics was a dry-lab based research project focused at solving complex biological problems through hypothesis testing, statistical analysis and the computer programming language Python. I was gobsmacked by the aspect of combining mathematics, computer science and biology, all into a single project to focus on an area of research. I could already sense the monotony from the other topics since they had an entirely different connotation in comparison to 'Genomics.' Prior to meeting my supervisor, funnily enough, the only aspect of bioinformatics or computational biology that seemed familiar was 'biology.' However, unlike students who appeared to be drowning in hopelessness at the idea of delving into research, I chose my project without any reluctance. Since the research project was part of a course-based curriculum consisting of a group of 7 final year undergraduates, a central theme provided by Dr Maurice HT Ling made it easy for all the students to compare and contrast their respective projects. Such a collaborative learning environment not only aided in a better understanding of the scientific process but also made the research experience available to a maximum number of students. This was brought about primarily by the integration of knowledge and inquiry, which is foundational to both research and critical thinking (Ramos Goyette and DeLuca, 2007, Linn et al., 2006, Temple et al., 2010, Shaffer et al., 2014, Auchincloss et al., 2014, Staub et al., 2016).

My project mainly dealt with the investigation of codon usage bias patterns, guanine-cytosine content, peptide distribution, amino acid and codon count variations in the *Pseudomonas balearica* DSM 6083^T genome. The reason behind choosing *P. balearica* firstly, was the absence of an all-inclusive research (less than 20 citations identified in PubMed for search term: ("pseudomonas" [MeSH Terms] OR "pseudomonas" [All Fields]) AND balearica[All Fields] till date) and secondly its potential industrial application in bioremediation, bioaugmentation and possible turnover of thiosulphate disproportionation. Due to the time constraints accompanying a three-month long project, an absolute conclusion could not be accomplished, however, the project undertaken instilled in me not only an appreciation of the process of scientific inquiry but also the mentor-mentee relationship. Weeks after the completion of my degree, I enquired Maurice about the steps that I needed to take to further develop my skills in bioinformatics. He suggested me to consider an apprenticeship. I therefore heeded his suggestion and tried to persuade him to take me under his tutelage, and to which he fortunately agreed. So far, I would highlight my acquaintance with Maurice as the most influential event in my academic life. Two of his qualities which I came to admire are his pedantries and approachability: his attention to detail would eventually motivate me to chip out my rough edges both in research and life.

My Reflections

As I recount all of my experiences in Singapore and reiterate in memory, I cannot help but recall an excerpt from Robert Frost's *The Road Not Taken* (1916), which has stuck with me since childhood –

> "Two roads diverged in a wood, and I-
> I took the one less travelled by,
> And that has made all the difference."

Being a contrarian, I believe that my journey up until now, albeit figuratively, emulates that of Frost's, and I am blessed to experience the 'difference' that it has made. Thus, it is imperative for me to overstate the instrumental contribution that Singapore has made to my life. Singapore's significant advancement in biomedical research can be directly attributed to their positive adaptation of the Hawthorne effect, which simply refers to the altered behaviour of a subject based on their awareness under observation (McCambridge et al., 2014, Wickström and Bendix, 2000, Parsons, 1974). Hence, instead of cowering down and getting overshadowed in the presence of biomedical giants such as the US and the UK, Singapore invested in education to establish themselves as the bioinformatics hub of the Asia-Pacific (Van Epps, 2006, Eisenhaber et al., 2009, Cyranoski, 2001).

One of the first things I realised when I returned to India was that unlike Singapore where students are exposed to scientific inquiry as early as high school, Indian undergraduates engross themselves in research not until late into their careers and possess little to no experience on the subtle art of scientific writing and plagiarism (Juyal et al., 2015, Misra et al., 2017). The reason I hereby allude to would be the complete textbook-based approach used by the Indian education system, which lacks a research component. I agree with both (Shyam, 2017) and (Balachandar, 2018): what Indian undergraduates lack is not intellectual capacity or the will to act, rather they lack direction and this lack of direction is the mainstay of the absence of a research culture in India. Similar in that respect, for the first 24 years of my life, I was a generic thinker whereas the succeeding years spent studying abroad taught me to think progressively with a specific purpose. I was analogous to a computer that understands only binary (the language that a computer speaks and is limited to in the form of 0s and 1s), accomplishing tasks at a generic-low level, however, throughout the course of my degree, I was able to abstract those generic components and comprehend it at a progressively higher level, thus bridging the gap between knowledge and applying that knowledge to achieve a targeted goal. Of course, this would be well-nigh impossible without Maurice who succinctly declared his expectations at the commencement of the project - *"Our eventual goal is to make publish worthy material."* This created a propensity in me to be highly conscious of the literature investigation and the experimental process and leave no stone unturned; from logbook keeping to the formulation of the entire project report, I learned to be more scrupulous. Throughout history, from ancient Greece to contemporary society, not only is the importance of the mentor-mentee relationship well established (Masters and Kreeger, 2017, Bauça, 2018, Rolfe, 2016, Pfund et al., 2016, Eller et al., 2014, Straus et al., 2013) but it has also been noted that the mentor matters more than the thesis project in question (Toklu and Fuller, 2017, Collins and Oliver, 2017, Ling, 2017). Similarly, the accomplishment of the research project was of little importance to me, relative to the relationship that I had formed with Maurice. For me, he became a role model who set the bar so high that there was a certain amount of dedication needed on my behalf to touch the bar, let alone raise it. I was studying about computer programming, statistical testing and bacterial genomic landscapes in addition to managing two clinical pathology portfolios and two clinical immunology reports, all of this in a span of three months; the honours year was truly intense. However, the *modus operandi* that I adopted from Maurice allowed me to accomplish all of my tasks in due time, and thus made the whole process less stressful.

Throughout my studies in Singapore, there were several difficulties (both financial and study-related); in these moments, I always remained calm and composed as I thought of my parents who had sacrificed tremendously to help me achieve a better education, and so I endured and survived. Although India promotes 'unity in diversity,' witnessing it in practice for all these years was truly a life-altering experience. Despite my intense liking for Singapore and its cuisine, which is a perfect blend of all Southeast Asian and Western food, I never liked its weather, which is almost synonymous to stagnant water. However, if I were to advise someone who wants to choose a similar path, I would definitely recommend so, but with certain caveats. In agreement with (Kennedy, 2015), the three 'must-have' qualities that I believe will be crucial to the individual are – firstly, an ability to embrace failure (as it is the most important aspect of the scientific process), secondly, an unquenchable thirst for knowledge (for the more you seek, the more you understand) and lastly, humility. If the individual is sans all three, he or she must be modest enough and be willing to learn in order to appreciate scientific inquiry because the combination of these three will yield a harmonious passion, which has greater importance than the attainment of good grades. In addition to this, adaptation (embracing change), multitasking (handling multiple projects synchronously), self-scrutinization (learning from mistakes and rectifying it) and confrontation of difficulties will constitute the one-page checklist for someone in my shoes. Hence, I am immensely grateful to my parents for allowing me to study in Singapore, which not only made me more appreciative of the world and the people in it but also filled my psychological void by guiding me with a clear vision of what to do with my life.

So, will I undertake such an endeavour again? My answer to that would be an emphatic yes. I would love to continue my postgraduate studies and attain the highest university degree in Singapore. I believe that its ever-extending scope of bioinformatics and computational biology will equip me with an unparalleled arsenal of tools that will further allow me to implement my newfound knowledge not only in translational research (Ranganathan et al., 2012, Kuznetsov et al., 2013, Fu and Lin, 2017) but also in answering some of the most sought after questions in human biology, many of which are reviewed in (Dev, 2015). I have a question of my own, albeit in infancy - *"What is the possible limit or capacity or threshold of cognition or more simplistically, of understanding (in terms of computation)?"* An answer to this, without a doubt, can provide a better understanding of both cognitive behaviour and function. Furthermore, as cited in (Ling, 2017) and originally proclaimed by Dobzhansky, I too believe that everything in biology will eventually become crystal clear in the light of evolution (Dobzhansky, 1973). Similar to Teresa Munro, a nursing student who took the opportunity to spend her sophomore year abroad in Spain to gain research experience, I truly believe that *"Studying abroad stays with you"* (Munro, 2017).

REFERENCES

Adams, D. R., Meyers, S. A. and Beidas, R. S. (2016) 'The relationship between financial strain, perceived stress, psychological symptoms, and academic and social integration in undergraduate students,' *Journal of American College Health* 64(5), pp. 362-370.

Auchincloss, L. C., Laursen, S. L., Branchaw, J. L., Eagan, K., Graham, M., Hanauer, D. I., Lawrie, G., McLinn, C. M., Pelaez, N., Rowland, S., Towns, M., Trautmann, N. M., Varma-Nelson, P., Weston, T. J. and Dolan, E. L. (2014) 'Assessment of course-based undergraduate research experiences: a meeting report,' *CBE life sciences education,* 13(1), pp. 29-40.

Augustine, J. (2017) 'Increased Educational Attainment among U.S. Mothers and their Children's Academic Expectations,' *Research in social stratification and mobility,* 52, pp. 15-25.

Back to the Future Part III, 1990. Directed by Zemeckis, R. United States of America: Universal Pictures.

Balachandar, G. D. (2018) 'Why do we lack a research culture? Analyzing the Indian Medical landscape - Response,' *Journal of orthopaedic case reports,* 8(2), pp. 110-110.

Bauça, J. M. (2018) 'Reflections on the Mentor-Mentee Relationship: A Symbiosis,' *EJIFCC,* 29(3), pp. 230-233.

Bogg, T. (2016) 'Self-control, dietary quality and new frontiers in the study of traits and wellness: A commentary on Keller, Hartmann and Siegrist,' *Psychology & Health,* 31(11), pp. 1328-1331.

Briki, W. (2017) 'Passion, Trait Self-Control, and Wellbeing: Comparing Two Mediation Models Predicting Wellbeing,' *Frontiers in psychology,* 8, pp. 841-841.

Carneiro, P., Meghir, C. and Parey, M. (2013) 'Maternal education, home environments, and the development of children and adolescents.,' *Journal of the European Economic Association,* 11(s1), pp. 123-160.

Chuan, Y. K. (2007) 'Singapore's biomedical sciences landscape,' *Biotechnology Journal,* 2(11), pp. 1331-1332.

Collins, K. and Oliver, S. W. (2017) 'Mentoring: what matters most?,' *The Clinical Teacher,* 14(4), pp. 298-300.

Cyranoski, D. (2001) 'Singapore invests in bioinformatics,' *Nature,* 410, pp. 293.

Dev, S. B. (2015) 'Unsolved problems in biology—The state of current thinking,' *Progress in Biophysics and Molecular Biology,* 117(2), pp. 232-239.

Dickson, M., Gregg, P. and Robinson, H. (2016) 'Early, late of never? When does parental education impact child outcomes?,' *Economic journal (London, England),* 126, pp. F184-F231.

Dobzhansky, T. (1973) 'Nothing in Biology Makes Sense except in the Light of Evolution,' *The American Biology Teacher,* 35(3), pp. 125-129.

Edgeworth, M. (1827) *Harry and Lucy concluded: being the last part of early lessons. In four volumes.* London: Printed for R. Hunter [and three others].

Eisenhaber, F., Kwoh, C. K., Ng, S. K., Sung, W. K. and Wong, L. (2009) 'Brief overview of bioinformatics activities in Singapore,' *PLoS computational biology,* 5(9), pp. e1000508-e1000508.

Eller, L. S., Lev, E. L. and Feurer, A. (2014) 'Key components of an effective mentoring relationship: a qualitative study,' *Nurse education today,* 34(5), pp. 815-820.

Freed, D. L. J. (2004) 'Doctors are not scientists but we still need science,' *BMJ (Clinical research ed.)*, 329(7460), pp. 294-294.

Fu, Z. and Lin, J. (2017) 'An Overview of Bioinformatics Tools and Resources in Allergy,' in Lin, J. & Alcocer, M. (eds.) *Food Allergens: Methods and Protocols*. New York, NY: Springer New York, pp. 223-245.

Goel, S., Angeli, F., Dhirar, N., Singla, N. and Ruwaard, D. (2018) 'What motivates medical students to select medical studies: a systematic literature review,' *BMC medical education*, 18(1), pp. 16-16.

Harackiewicz, J. M., Canning, E. A., Tibbetts, Y., Giffen, C. J., Blair, S. S., Rouse, D. I. and Hyde, J. S. (2014) 'Closing the Social Class Achievement Gap for First-Generation Students in Undergraduate Biology,' *Journal of educational psychology*, 106(2), pp. 375-389.

Heikkilä, T. J., Hyppölä, H., Vänskä, J., Aine, T., Halila, H., Kujala, S., Virjo, I., Sumanen, M. and Mattila, K. (2015) 'Factors important in the choice of a medical career: a Finnish national study,' *BMC medical education*, 15, pp. 169-169.

Higgins, E. T., Marguc, J. and Scholer, A. A. (2012) 'Value from Adversity: How We Deal with Adversity Matters,' *Journal of experimental social psychology*, 48(4), pp. 965-967.

Howell, J. (1659) *Paroimiographia Proverbs, or, Old sayed savves & adages in English (or the Saxon toung), Italian, French, and Spanish, whereunto the British for their great antiquity and weight are added. Proverbs, or, Old sayed savves & adages*. London: Grismond, John. Available at: http://name.umdl.umich.edu/A44738.0001.001.

Janke, S., Rudert, S. C., Marksteiner, T. and Dickhäuser, O. (2017) 'Knowing One's Place: Parental Educational Background Influences Social Identification with Academia, Test Anxiety, and Satisfaction with Studying at University,' *Frontiers in psychology*, 8, pp. 1326-1326.

Jury, M., Smeding, A. and Darnon, C. (2015) 'First-generation students' underperformance at university: the impact of the function of selection,' *Frontiers in psychology*, 6, pp. 710-710.

Juyal, D., Thawani, V. and Thaledi, S. (2015) 'Rise of academic plagiarism in India: Reasons, solutions and resolution,' *Lung India: official organ of Indian Chest Society*, 32(5), pp. 542-543.

Kalra, B., Joshi, A., Kalra, S., Shanbhag, V. G., Kunwar, J., Singh Balhara, Y. P., Chaudhary, S., Khandelwal, D., Aggarwal, S., Priya, G., Verma, K., Baruah, M. P., Sahay, R., Bajaj, S., Agrawal, N., Pathmanathan, S., Prasad, I., Chakraborty, A. and Ram, N. (2018) 'Coping with Illness: Insight from the Bhagavad Gita,' *Indian journal of endocrinology and metabolism*, 22(4), pp. 560-564.

Kalra, S., Joshi, A., Kalra, B., Shanbhag, V. G., Bhattacharya, R., Verma, K., Baruah, M. P., Sahay, R., Bajaj, S., Agrawal, N., Chakraborty, A., Balhara, Y. P. S., Chaudhary, S., Khandelwal, D., Aggarwal, S., Ram, N., Jacob, J., Julka, S., Priya, G., Bhattacharya, S. and Dalal, K. (2017) 'Bhagavad Gita for the Physician,' *Indian journal of endocrinology and metabolism*, 21(6), pp. 893-897.

Kennedy, P. G. E. (2015) 'My life as a clinician-scientist: trying to bridge the perceived gap between medicine and science,' *DNA and cell biology*, 34(6), pp. 383-390.

Kuznetsov, V., Lee, H. K., Maurer-Stroh, S., Molnár, M. J., Pongor, S., Eisenhaber, B. and Eisenhaber, F. (2013) 'How bioinformatics influences health informatics: usage of biomolecular sequences, expression profiles and automated microscopic image analyses for clinical needs and public health,' *Health information science and systems*, 1, pp. 2-2.

Ling, M. H. T. (2017) 'A personal narrative of 6 pre-university research projects over 7 years (2009-2015) yielding 19 manuscripts,' *MOJ Proteomics & Bioinformatics*, 6(3), pp. 270-281.

Linn, M. C., Lee, H. S., Tinker, R., Husic, F. and Chiu, J. L. (2006) 'Teaching and Assessing Knowledge Integration in Science,' *Science*, 313(5790), pp. 1049.

Lum, L. H. W., Poh, K. K. and Tambyah, P. A. (2018) 'Winds of change in medical education in Singapore: what does the future hold?' *Singapore medical journal*, 59(12), pp. 614-615.

Masters, K. S. and Kreeger, P. K. (2017) 'Ten simple rules for developing a mentor-mentee expectations document,' *PLoS computational biology,* 13(9), pp. e1005709-e1005709.

McCambridge, J., Witton, J. and Elbourne, D. R. (2014) 'Systematic review of the Hawthorne effect: new concepts are needed to study research participation effects,' *Journal of clinical epidemiology,* 67(3), pp. 267-277.

Miri, M. R., Kermani, T., Khoshbakht, H. and Moodi, M. (2013) 'The relationship between emotional intelligence and academic stress in students of medical sciences,' *Journal of education and health promotion,* 2, pp. 40-40.

Misra, D. P., Ravindran, V., Wakhlu, A., Sharma, A., Agarwal, V. and Negi, V. S. (2017) 'Plagiarism: a Viewpoint from India,' *Journal of Korean medical science,* 32(11), pp. 1734-1735.

Mohd Matore, M. E. E., Khairani, A. Z. and Razak, N. A. (2015) 'The Influence of AQ on the Academic Achievement among Malaysian Polytechnic Students,' *International Education Studies,* 8(6), pp. 69-74.

Munro, T. (2017) 'Studying abroad stays with you,' *Nursing,* 47(2), pp. 69-70.

Nath, I. 'New technologies for health'. *New Technologies session. Commonwealth Conference 2017,* Singapore.

Nath, I., Saini, C. and Valluri, V. L. (2015) 'Immunology of leprosy and diagnostic challenges,' *Clinics in Dermatology,* 33(1), pp. 90-98.

Nature (2010) 'Singapore's salad days are over,' *Nature,* 468, pp. 731.

Parsons, H. M. (1974) 'What Happened at Hawthorne?,' *Science,* 183(4128), pp. 922.

Pfund, C., Byars-Winston, A., Branchaw, J., Hurtado, S. and Eagan, K. (2016) 'Defining Attributes and Metrics of Effective Research Mentoring Relationships,' *AIDS and behavior,* 20 Suppl 2(Suppl 2), pp. 238-248.

Pufall, E., Eaton, J. W., Nyamukapa, C., Schur, N., Takaruza, A. and Gregson, S. (2016) 'The relationship between parental education and children's schooling in a time of economic turmoil: The case of East

Zimbabwe, 2001 to 2011,' *International journal of educational development,* 51, pp. 125-134.

Ramos Goyette, S. and DeLuca, J. (2007) 'A semester-long student-directed research project involving enzyme immunoassay: appropriate for immunology, endocrinology, or neuroscience courses,' *CBE life sciences education,* 6(4), pp. 332-342.

Ranasinghe, P., Wathurapatha, W. S., Mathangasinghe, Y. and Ponnamperuma, G. (2017) 'Emotional intelligence, perceived stress and academic performance of Sri Lankan medical undergraduates,' *BMC medical education,* 17(1), pp. 41-41.

Ranganathan, S., Tongsima, S., Chan, J., Tan, T. W. and Schönbach, C. (2012) 'Advances in translational bioinformatics and population genomics in the Asia-Pacific,' *BMC genomics,* 13 Suppl 7(Suppl 7), pp. S1-S1.

Rolfe, A. (2016) 'The mentor's role,' *Korean journal of medical education,* 28(3), pp. 315-316.

Samarasekera, D. D., Ooi, S., Yeo, S. P. and Hooi, S. C. (2015) 'Medical education in Singapore,' *Medical Teacher,* 37(8), pp. 707-713.

Shaffer, C. D., Alvarez, C. J., Bednarski, A. E., Dunbar, D., Goodman, A. L., Reinke, C., Rosenwald, A. G., Wolyniak, M. J., Bailey, C., Barnard, D., Bazinet, C., Beach, D. L., Bedard, J. E. J., Bhalla, S., Braverman, J., Burg, M., Chandrasekaran, V., Chung, H.-M., Clase, K., Dejong, R. J., Diangelo, J. R., Du, C., Eckdahl, T. T., Eisler, H., Emerson, J. A., Frary, A., Frohlich, D., Gosser, Y., Govind, S., Haberman, A., Hark, A. T., Hauser, C., Hoogewerf, A., Hoopes, L. L. M., Howell, C. E., Johnson, D., Jones, C. J., Kadlec, L., Kaehler, M., Silver Key, S. C., Kleinschmit, A., Kokan, N. P., Kopp, O., Kuleck, G., Leatherman, J., Lopilato, J., Mackinnon, C., Martinez-Cruzado, J. C., McNeil, G., Mel, S., Mistry, H., Nagengast, A., Overvoorde, P., Paetkau, D. W., Parrish, S., Peterson, C. N., Preuss, M., Reed, L. K., Revie, D., Robic, S., Roecklein-Canfield, J., Rubin, M. R., Saville, K., Schroeder, S., Sharif, K., Shaw, M., Skuse, G., Smith, C. D., Smith, M. A., Smith, S. T., Spana, E., Spratt, M., Sreenivasan, A., Stamm, J., Szauter, P., Thompson, J. S., Wawersik, M., Youngblom, J., Zhou, L.,

Mardis, E. R., Buhler, J., Leung, W., Lopatto, D. and Elgin, S. C. R. (2014) 'A course-based research experience: how benefits change with increased investment in instructional time,' *CBE life sciences education,* 13(1), pp. 111-130.

Shyam, A. D. (2017) 'Why do we lack a Research Culture? Analysing the Indian Medical Landscape,' *Journal of orthopaedic case reports,* 7(5), pp. 1-2.

Smith, R. (2004) 'Doctors are not scientists,' *BMJ,* 328(7454).

Staub, N. L., Poxleitner, M., Braley, A., Smith-Flores, H., Pribbenow, C. M., Jaworski, L., Lopatto, D. and Anders, K. R. (2016) 'Scaling Up: Adapting a Phage-Hunting Course to Increase Participation of First-Year Students in Research,' *CBE life sciences education,* 15(2), pp. ar13.

Straus, S. E., Johnson, M. O., Marquez, C. and Feldman, M. D. (2013) 'Characteristics of successful and failed mentoring relationships: a qualitative study across two academic health centers,' *Academic medicine: journal of the Association of American Medical Colleges,* 88(1), pp. 82-89.

Temple, L., Cresawn, S. G. and Monroe, J. D. (2010) 'Genomics and bioinformatics in undergraduate curricula: Contexts for hybrid laboratory/lecture courses for entering and advanced science students,' *Biochemistry and Molecular Biology Education,* 38(1), pp. 23-28.

The Bhagavad-Gita. (1929): Chicago, Ill.: The University of Chicago press, [1929] [©1929].

Tibbetts, Y., Priniski, S. J., Hecht, C. A., Borman, G. D. and Harackiewicz, J. M. (2018) 'Different Institutions and Different Values: Exploring First-Generation Student Fit at 2-Year Colleges,' *Frontiers in psychology,* 9, pp. 502-502.

Toklu, H. Z. and Fuller, J. C. (2017) 'Mentor-mentee Relationship: A Win-Win Contract in Graduate Medical Education,' *Cureus,* 9(12), pp. e1908-e1908.

Vallerand, R. J. (2012) 'The role of passion in sustainable psychological well-being,' *Psychology of Well-Being: Theory, Research and Practice,* 2(1), pp. 1.

Van Epps, H. L. (2006) 'Singapore's multibillion dollar gamble,' *The Journal of experimental medicine,* 203(5), pp. 1139-1142.

Verma, S., Aggarwal, A. and Bansal, H. (2017) *The Relationship between Emotional Intelligence (EQ) and Adversity Quotient (AQ).*

Verner-Filion, J., Vallerand, R. J., Amiot, C. E. and Mocanu, I. (2017) 'The two roads from passion to sport performance and psychological well-being: The mediating role of need satisfaction, deliberate practice, and achievement goals,' *Psychology of Sport and Exercise,* 30, pp. 19-29.

Wickström, G. and Bendix, T. (2000) 'The "Hawthorne effect" - what did the original Hawthorne studies actually show?,' *Scandinavian Journal of Work, Environment & Health,* (4), pp. 363-367.

Wong, J. (2017) 'Next Frontier: Collaborative Partnerships and Digitalisation,' *Horizons,* (no. 4), pp. 19.

Woodward, A., Thomas, S., Jalloh, M. B., Rees, J. and Leather, A. (2017) 'Reasons to pursue a career in medicine: a qualitative study in Sierra Leone,' *Global health research and policy,* 2, pp. 34-34.

Zamri Khairani, A. and Syed Abdullah, S. M. (2018) 'Relationship between adversity quotient and academic well-being among Malaysian undergraduates,' *Asian Journal of Scientific Research,* 11(1), pp. 51-55.

Date Submitted: February 24, 2019. Date Accepted: March 13, 2019.

In: Current STEM. Volume 2
Editor: Maurice HT Ling

ISBN: 978-1-53616-042-0
© 2019 Nova Science Publishers, Inc.

Chapter 3

SEcured REcorder BOx (SEREBO) Version 1.0

Maurice HT Ling[*]

Colossus Technologies LLP, Singapore

HOHY PTE LTD, Singapore

Abstract

Data authenticity is crucial in many industries. A major aspect of data authenticity is to ensure that a created file is not fraudulently or purposefully edited; for example, changing the data file without affecting the date time stamp. Blockchain technology ensures data authenticity as recorded data is not mutable. This manuscript documents the implementation and the codes of SEREBO and licensed under GNU General Public License version 3. SEREBO codebase is hosted and available for forking at https://github.com/mauriceling/serebo.

Keywords: data management, research records, blockchain, immutability

---

[*] Corresponding Author's E-mail: mauriceling@acm.org, mauriceling@colossus-tech.com.

PROBLEM SCENARIO AND INTRODUCTION

One of the hallmarks of biological and biomedical research today is data; more specifically, big data (Dolinski and Troyanskaya, 2015; Suwinski et al., 2019). As a bioinformatics researcher, I often generate data files as I go about my work. Take for example, an Excel file, `sampleTimeSeries.xlsx`, that I had generated on 15th May 2015. Three years later, on 10th June 2018, how can I demonstrate that `sampleTimeSeries.xlsx` had not been changed/edited since 15th May 2015? If I had the intention to change the file on 10th December 2016, I can safely set my computer clock back to 15th May 2015, change the file, and the date will still be 15th May 2015. A recent study (Bik et al., 2018) analyzing 960 published papers found 59 (6.1%) papers contained inappropriately duplicated images, resulting in 5 retractions.

What if there is a way for me to log my file into a system, on 15th May 2015, and this record (the log statement) is not editable? To ensure that this record is not editable, it can be put on a blockchain. One of the most useful features of a blockchain is resistance against modification of data (Zheng et al., 2017). This resulted in the implementation of system called SEREBO – SEcured REcorder BOx (Ling, 2018a) – comprising of SEREBO Black Box and SEREBO Notary. This manuscript documents the implementation of SEREBO. SEREBO is available for forking at https://github.com/mauriceling/serebo under GNU General Public License version 3 for non-commercial or academic use only.

SEREBO Black Box is inspired by the black boxes (cockpit voice recorder and flight data recorder) in airliners (Pierce, 2010). The intended purpose is to track and audit research records under the following premise – Given a set of data files, is there a system to log and verify that these files had not been changed or edited since its supposed creation? SEREBO Black Box addresses this issue by three approaches. Firstly, the data files can be used to generate a file hash as an edit in the file will result in a different hash. As such, a file that generates the same hash across two different points in time can be safely assumed unedited during this time span. Secondly, the file hash is securely recorded with amendment protection. SEREBO Black Box records the hash and registers the hash into a blockchain. The main concept of blockchain is that the hash of previous (parent) block is concatenated with the data (file hash in this case) of the current block to generate a hash for the current block. Therefore, any amendments in earlier blocks can be easily detected as the blockchain grows as only amendments to the latest block cannot be detected. This property makes the data in blockchain immutable once it is locked within a chain (Zheng et al., 2017). Lastly, SEREBO Notary is implemented as a web-based notarized by one or more independent notary to notarize SEREBO Black Boxes, which adds another layer of modification restriction to downstream SEREBO Black Boxes. The architecture of SEREBO Notary is based on a previous work, NotaLogger (Ling, 2013).

ARCHITECTURE AND IMPLEMENTATION

SEREBO consists of three components (Figure 1):

- SEREBO Command Line, which is the command-line user interface (CLI) to access both SEREBO Black Box and SEREBO Notary. This is implemented in `serebo.py` (main command-line processor) and `serebo_notary_api.py` (interface to access SEREBO Notary) files.
- SEREBO Black Box, which is the blockchain-based data storage facility. This is implemented in `serebo_api.py` (interface between SEREBO Black Box and the external world) and `sereboDB.py` (talks to SEREBO Black Box database) files.
- SEREBO Notary, which acts as a public and independent notary to sign off (notarize) SEREBO Black Boxes. This is implemented in `services.py` (interface for SEREBO Notary to other SEREBO Black Boxes) and `serebo_notabase.py` (talks to SEREBO Notary database) files.

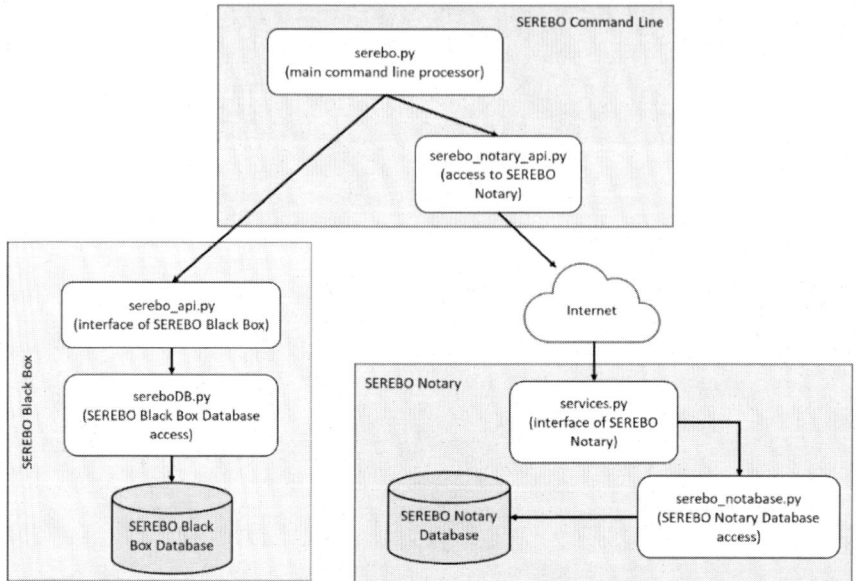

Figure 1. Overview of SEREBO.

The entry point to both SEREBO Black Box and SEREBO Notary is SEREBO Command Line, implemented using Python 3 and Python-Fire module (https://github.com/google/python-fire), which aims to simplify the implementation of command-line interface in Python 3. The means of access is using `serebo.py` file. At the command line level, the general syntax is `python serebo.py [operation] [option(s)]`. The operations/commands in SEREBO Command Line can be classified into three categories, consisting of a total of 32 operations, as follows:

1. SEREBO Black Box Operations
    1.1. `backup`: Backup SEREBO Black Box
    1.2. `dump`: Dump out data (text backup) from SEREBO Black Box
    1.3. `fhash`: Generate and print out hash of a file
    1.4. `init`: Initialize SEREBO Black Box
    1.5. `intext`: Insert a text string into SEREBO Black Box
    1.6. `localcode`: Generate a random string, and log this generation into SEREBO Black Box

1.7. `localdts`: Get date time string
1.8. `logfile`: Log a file into SEREBO Black Box
1.9. `ntpsign`: Self-sign (self-notarization) SEREBO Black Box using NTP (Network Time Protocol) Server
1.10. `searchmsg`: Search SEREBO Black Box data log for a message
1.11. `searchdesc`: Search SEREBO Black Box data log for a description
1.12. `searchfile`: Search SEREBO Black Box data log for a file log
1.13. `selfsign`: Self-sign (self-notarization) SEREBO Black Box
1.14. `shash`: Generate hash for a data string using SEREBO Black Box
1.15. `sysdata`: Print out data and test hashes of current platform
1.16. `sysrecord`: Record data and test hashes of current platform into SEREBO Black Box
1.17. `viewntpnote`: View all self-notarization(s) by NTP time server for this SEREBO Black Box
1.18. `viewselfnote`: View self-notarization(s) for current SEREBO Black Box
2. SEREBO Notary Operations
    1.1. `changealias`: Change alias for a specific SEREBO Notary registration
    1.2. `notarizebb`: Notarize SEREBO Black Box with SEREBO Notary
    1.3. `register`: Register SEREBO Black Box with SEREBO Notary
    1.4. `viewsnnote`: View notarization(s) by SEREBO Notary(ies) for current SEREBO Black Box
    1.5. `viewreg`: View current SEREBO Notary registrations for current SEREBO Black Box

3. Audit Operations
    1.1. `audit_blockchainflow`: Trace the decendency of blockchain records (also known as blocks) within SEREBO Black Box
    1.2. `audit_blockchainhash`: Check for accuracy in blockchain hash generation within SEREBO Black Box
    1.3. `audit_count`: Check for equal numbers of records in data log and blockchain in SEREBO Black Box
    1.4. `audit_data_blockchain`: Check for accuracy in data log and blockchain mapping in SEREBO Black Box
    1.5. `audit_datahash`: Check for accuracy of hash generations in data log within SEREBO Black Box
    1.6. `audit_notarizebb`: Check for SEREBO Black Box notarization records in SEREBO Notary
    1.7. `audit_register`: Check for registration between SEREBO Black Box and SEREBO Notary
    1.8. `checkhash`: Compare record hash from SEREBO Black Box with that in a file
    1.9. `dumphash`: Dump out record hash from SEREBO Black Box into a file

The black box is implemented as a SQLite database consisting of 7 tables (defined in sereboDB.py):

1. `metadata` table stores information about the SEREBO Black Box. At creation using `init` command, date time stamp of creation and a randomly generated 512-character string to represent the identity of the SEREBO Black Box will be recorded.
2. `notary` table stores registration record(s) between SEREBO Black Box to one or more SEREBO Notary(ies), registered using register command.
3. `systemdata` table stores data and test hashes of current platform using sysrecord command, which is used to provide a data baseline

and for future checking for potential differences in processing outcomes due to different platforms.
4. `datalog` table stores the actual data to be logged and its corresponding hash.
5. `blockchain` table is the backbone of SEREBO Black Box, where each record/tuple represents a block in the blockchain.
6. `eventlog` table stores audit trail of the data logging event when a piece of data is stored in datalog table.
7. `eventlog_datamap` table provides the second audit trail of data hash and block chain hashes when a piece of data is stored in datalog table.

Input into SEREBO Black Box, via SEREBO Command Line, can either be a string or a file. When the input is a string, the data is the actual input string and the description can be used to provide additional information about the string. When a file is given, a series of 12 hashes is generated from the file (Ling, 2018b) to reduce the possibility of hash collision (Boneh and Boyen, 2006; Broder and Mitzenmacher, 2001; Rasjid et al., 2017). The series of hashes is then concatenated to form the data. The absolute and relative file path are added to the description. Hence, both file and string input are reduced to a data component and a description component (Figure 2), which is then used to generate a series of 12 hashes, called DataHash, after the inclusion of a date time stamp for the input event. The data, description, date time stamp, and DataHash are recorded in the datalog table.

The latest block in blockchain table is identified and used as the parental block. Three attributes/values from the parental block are extracted – date time stamp (known as parental Date Time stamp), random string (a 32-character random string, known as parental Random String), and block hash (known as parental BlockHash). These 3 parental attributes will be combined with DataHash to generate a BlockHash (known as current BlockHash). A new 32-character Random String (known as current Random String) will be generated and a new block is generated (and will be used as parental block for the next succeeding block), which consists of the following attributes:

1. Date time stamp of the data entry event (identical to date time stamp in the datalog table)
2. Current Random String
3. Current BlockHash
4. Parental block ID
5. Parental Date Time stamp
6. Parental Random String
7. Parental BlockHash
8. Data (which is DataHash)

Finally, the description and date time stamp are logged in eventlog table. This is followed by logging current BlockHash, parental BlockHash, and DataHash are logged in eventlog_datamap table.

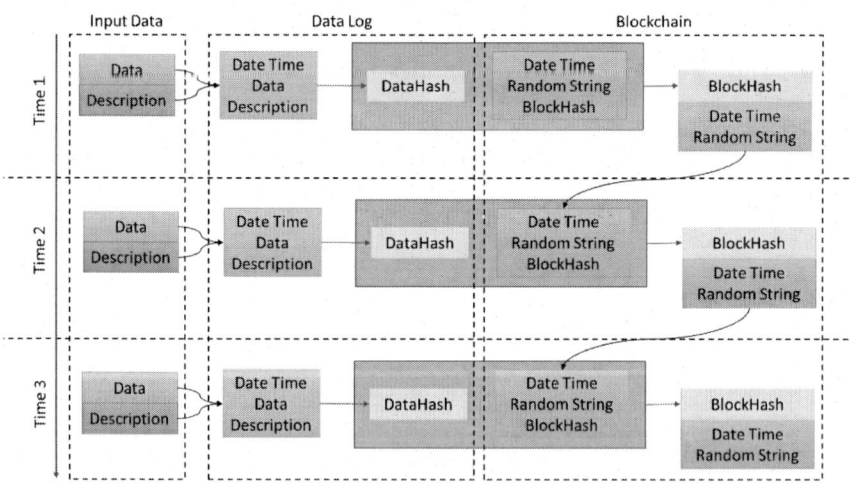

Figure 2. Overview of Operations in SEREBO Black Box [Adapted from (Ling, 2018a)).

SEREBO Notary is a web application built on Web2Py framework (Pierro, 2008) and exposes a set of XMLRPC web services to provide an independent agent/platform as a notary service. The architecture of SEREBO Notary is based on a previous work, NotaLogger (Ling, 2013). A SEREBO Black Box can be registered with one or more SEREBO Notaries. An instance SEREBO Notary has been set up for public use and is accessible at https://mauricelab.pythonanywhere.com/ serebo_notary/ services/call/xmlrpc. After registration, a SEREBO Notary can be called to notarize the SEREBO Black Box. During which, the SEREBO Black Box will generate a local date time stamp (known as Black Box date time stamp) and 32-character random string (known as Black Box code) to be transmitted to the SEREBO Notary, together with identity of the SEREBO Black Box. Upon receiving this set of information, the SEREBO Notary will generate a date time stamp (known as Notary date time stamp) and a 32-character random string (known as Notary code). In addition, the SEREBO Notary will generate a set of hashes (known as common code) from the Black Box code and Notary code. All information will be logged in SEREBO Notary before transmitting Notary date time stamp, Notary code, and common code back to the requesting SEREBO Black Box for logging into datalog table, which will in turn trigger the logging into blockchain.

CODE FILES FOR SEREBO COMMAND LINE[1]

File Name: SEREBO.PY

```
'''!
    Secured Recorder Box (SEREBO) Command Line Interface
(CLI)

    Date created: 17th May 2018
    License: GNU General Public License version 3 for
academic or not-for-profit use only
```

---

[1] These code files had been formatted for display and printing purposes; hence, different from the original codes in http://github.com/mauriceling/serebo on a line-by-line basis. However, they are logically, syntactically, and semantically identical to the original codes.

SEREBO is free software: you can redistribute it and/or modify it under the terms of the GNU General Public License as published by the Free Software Foundation, either version 3 of the License, or (at your option) any later version.

This program is distributed in the hope that it will be useful, but WITHOUT ANY WARRANTY; without even the implied warranty ofMERCHANTABILITY or FITNESS FOR A PARTICULAR PURPOSE. See theGNU General Public License for more details.

You should have received a copy of the GNU General Public Licensealong with this program. If not, see <http://www.gnu.org/licenses/>.

```
'''
import random
import os
import sqlite3

import fire

import serebo_blackbox as bb
import serebo_notary_api as notary

def initialize(bbpath='serebo_blackbox\\blackbox.sdb'):
    '''!
    Function to initialize SEREBO blackbox.

    Usage:

        python serebo.py init
        --bbpath=<path to SEREBO black box>

    For example:

        python serebo.py init \
        --bbpath='serebo_blackbox\\blackbox.sdb'

    @param bbpath String: Path to SEREBO black box. Default = 'serebo_blackbox\\blackbox.sdb'.
    '''
```

```
   db = bb.connectDB(bbpath)
   try:
      sqlstmt = '''insert into metadata (key, value)
      values
         ('serebo_blackbox_path', '%s');''' %\
         (str(db.path))
      db.cur.execute(sqlstmt)
      db.conn.commit()
   except sqlite3.IntegrityError: pass
   print('')
   return {'SEREBO Black Box': db,
         'Black Box Path': str(db.path)}

def insertText(message, description='NA',
            bbpath='serebo_blackbox\\blackbox.sdb'):
   '''!
   Function to insert a text string into SEREBO blackbox.

   Usage:

      python serebo.py intext
      --message=<text message to be inserted>
      --description=<explanatory description for this
      insertion>
      --bbpath=<path to SEREBO black box>

   For example:

      python serebo.py intext \
      --message="This is a text message for insertion" \
      --description="Texting 1" \
      --bbpath='serebo_blackbox\\blackbox.sdb'

   @param message String: Text string to be inserted.
   @param description String: Explanation string for this
   entry event. Default = NA.
   @param bbpath String: Path to SEREBO black box. Default
   = 'serebo_blackbox\\blackbox.sdb'.
   '''
   db = bb.connectDB(bbpath)
```

```
    rdata = bb.insertText(db, message, description)
    print('')
    print('Insert Text Status ...')
    return {'SEREBO Black Box': db,
            'Black Box Path': str(db.path),
            'Date Time Stamp': str(rdata['DateTimeStamp']),
            'Message': str(rdata['Data']),
            'Description': str(rdata['UserDescription']),
            'Data Hash': str(rdata['DataHash'])}

def logFile(filepath, description='NA',
            bbpath='serebo_blackbox\\blackbox.sdb'):
    '''!
    Function to log a file into SEREBO blackbox.

    Usage:

        python serebo.py logfile
        --filepath=<path of file to log>
        --description=<explanatory description for this
        insertion>
--bbpath=<path to SEREBO black box>

    For example:

        python serebo.py logfile \
        --filepath=doxygen_serebo \
        --description="Doxygen file for SEREBO" \
        --bbpath='serebo_blackbox\\blackbox.sdb'

    @param fileapth String: Path of file to log in SEREBO
        black box.
    @param description String: Explanation string for this
        entry event. Default = NA.
    @param bbpath String: Path to SEREBO black box. Default
        = 'serebo_blackbox\\blackbox.sdb'.
    '''
    db = bb.connectDB(bbpath)
    rdata = bb.logFile(db, filepath, description)
    print('')
```

```python
    print('File Logging Status ...')
    return {'SEREBO Black Box': db,
            'Black Box Path': str(db.path),
            'Date Time Stamp': str(rdata['DateTimeStamp']),
            'File Hash': str(rdata['Data']),
            'Description': str(rdata['UserDescription']),
            'Data Hash': str(rdata['DataHash'])}

def systemData():
    '''
    Function to print out data and test hashes of current \
    platform - This does not insert a record into SEREBO
    Black
    Box.

    Usage:

        python serebo.py sysdata
    '''
    data = bb.systemData()
    print('')
    print('System Data ...')
    return {'architecture': str(data['architecture']),
            'machine': str(data['machine']),
            'node': str(data['node']),
            'platform': str(data['platform']),
            'processor': str(data['processor']),
            'python_build': str(data['python_build']),
            'python_compiler':
            str(data['python_compiler']),
            'python_implementation': \
                str(data['python_implementation']),
            'python_branch': str(data['python_branch']),
            'python_revision':
            str(data['python_revision']),
            'python_version': str(data['python_version']),
            'release': str(data['release']),
            'system': str(data['system']),
            'version': str(data['version']),
            'hashdata': str(data['hashdata']),
```

```
                'hash_md5': str(data['hash_md5']),
                'hash_sha1': str(data['hash_sha1']),
                'hash_sha224': str(data['hash_sha224']),
                'hash_sha3_224': str(data['hash_sha3_224']),
                'hash_sha256': str(data['hash_sha256']),
                'hash_sha3_256': str(data['hash_sha3_256']),
                'hash_sha384': str(data['hash_sha384']),
                'hash_sha3_384': str(data['hash_sha3_384']),
                'hash_sha512': str(data['hash_sha512']),
                'hash_sha3_512': str(data['hash_sha3_512']),
                'hash_blake2b': str(data['hash_blake2b']),
                'hash_blake2s': str(data['hash_blake2s'])}

def systemRecord(bbpath='serebo_blackbox\\blackbox.sdb'):
    '''!
    Function to record data and test hashes of current
    platform.

    Usage:

        python serebo.py sysrecord \
        --bbpath=<path to SEREBO black box>

    For example:

        python serebo.py sysrecord \
        --bbpath='serebo_blackbox\\blackbox.sdb'

    @param bbpath String: Path to SEREBO black box. Defaul
        = 'serebo_blackbox\\blackbox.sdb'.
    '''
    db = bb.connectDB(bbpath)
    data = bb.systemData()
    dtstamp = bb.dateTime(db)
    sqlstmt = '''insert into systemdata (dtstamp, key,
    value)
    values ('%s', '%s', '%s');'''
    print('')
    print('System Data ...')
    for k in data:
```

```
        if k != 'hashdata':
            db.cur.execute(sqlstmt % (str(dtstamp), str(k),
                                      str(data[k])))
    db.conn.commit()
    return {'SEREBO Black Box': db,
            'Black Box Path': str(db.path),
            'architecture': str(data['architecture']),
            'machine': str(data['machine']),
            'node': str(data['node']),
            'platform': str(data['platform']),
            'processor': str(data['processor']),
            'python_build': str(data['python_build']),
            'python_compiler':
            str(data['python_compiler']),
            'python_implementation': \
                str(data['python_implementation']),
            'python_branch': str(data['python_branch']),
            'python_revision':
            str(data['python_revision']),
            'python_version': str(data['python_version']),
            'release': str(data['release']),
            'system': str(data['system']),
            'version': str(data['version']),
            'hashdata': str(data['hashdata']),
            'hash_md5': str(data['hash_md5']),
            'hash_sha1': str(data['hash_sha1']),
            'hash_sha224': str(data['hash_sha224']),
            'hash_sha3_224': str(data['hash_sha3_224']),
            'hash_sha256': str(data['hash_sha256']),
            'hash_sha3_256': str(data['hash_sha3_256']),
            'hash_sha384': str(data['hash_sha384']),
            'hash_sha3_384': str(data['hash_sha3_384']),
            'hash_sha512': str(data['hash_sha512']),
            'hash_sha3_512': str(data['hash_sha3_512']),
            'hash_blake2b': str(data['hash_blake2b']),
            'hash_blake2s': str(data['hash_blake2s'])}

def fileHash(filepath):
    '''!
    Function to generate and print out hash of a file.
```

```
Usage:

    python serebo.py fhash \
    --filepath=<path of file to hash>

For example:

    python serebo.py fhash \
    --filepath=doxygen_serebo

@param fileapth String: Path of file to log in SEREBO
black box.
'''
fHash = bb.fileHash(filepath)
print('')
return {'File Path': str(filepath),
        'File Hash': str(fHash)}

def localCode(length, description=None,
              bbpath='serebo_blackbox\\blackbox.sdb'):
    '''!
    Function to generate a random string, and log this
    generation into SEREBO Black Box.

    Usage:

        python serebo.py localcode \
        --length=<length of random string>
        --description=<explanatory description for this
        insertion>
        --bbpath=<path to SEREBO black box>

    For example:

        python serebo.py localcode \
        --length=10 \
        --description="Notarizing certificate ABC123" \
        --bbpath='serebo_blackbox\\blackbox.sdb'
```

```
    @param length Integer: Length of random string to
    generate
    @param description String: Explanation string for this
    entry event. Default = None.
    @param bbpath String: Path to SEREBO black box. Default
    = 'serebo_blackbox\\blackbox.sdb'.
    '''
    db = bb.connectDB(bbpath)
    rstring = bb.randomString(db, length)
    description = ['Local random string generation'] + \
        [description]
    description = ' | '.join(description)
    rdata = bb.insertFText(db, rstring, description)
    print('')
    print('Generate Random String (Local) ...')
    return {'SEREBO Black Box': db,
            'Black Box Path': str(db.path),
            'Date Time Stamp': str(rdata['DateTimeStamp']),
            'Random String': str(rstring)}

def localDTS(bbpath='serebo_blackbox\\blackbox.sdb'):
    '''!
    Function to get date time string. This event is not
    logged.

    Usage:

        python serebo.py localdts \
        --bbpath=<path to SEREBO black box>

    For example:

        python serebo.py localdts \
        --bbpath='serebo_blackbox\\blackbox.sdb'

    @param bbpath String: Path to SEREBO black box. Default
    = 'serebo_blackbox\\blackbox.sdb'.
    '''
    db = bb.connectDB(bbpath)
    dts = bb.dateTime(db)
```

```
    print('')
    return {'SEREBO Black Box': db,
            'Black Box Path': str(db.path),
            'Date Time Stamp': str(dts)}

def stringHash(dstring,
               bbpath='serebo_blackbox\\blackbox.sdb'):
    '''!
    Function to generate hash for a data string. This event
    is not logged.

    Usage:

        python serebo.py shash
        --dstring=<string to hash>
        --bbpath=<path to SEREBO black box>

    For example:

        python serebo.py shash \
        --dstring="SEREBO is hosted at
        https://github.com/mauriceling/serebo" \
        --bbpath='serebo_blackbox\\blackbox.sdb'

    @param dstring String: String to generate hash.
    @param bbpath String: Path to SEREBO black box. Default
    = 'serebo_blackbox\\blackbox.sdb'.
    '''
    db = bb.connectDB(bbpath)
    x = bb.stringHash(db, dstring)
    print('')
    return {'SEREBO Black Box': db,
            'Black Box Path': str(db.path),
            'Data String': str(dstring),
            'Data Hash': str(x)}

def registerBlackbox(owner, email, alias,
                     notaryURL='https://mauricelab.
                     pythonanywhere.com/serebo_notary/
                     services/call/xmlrpc',
```

```
                bbpath='serebo_blackbox\\blackbox.sdb'):
'''
Function to register SEREBO Black Box with SEREBO
Notary.

Usage:

    python serebo.py register
    --alias=<alias for this SEREBO Notary>
    --notaryURL="https://mauricelab.pythonanywhere.
    com/serebo_notary/services/call/xmlrpc"
    --owner=<owner's name>
    --email=<owner's email>
    --bbpath=<path to SEREBO black box>

For example:

    python serebo.py register \
    --alias="NotaryPythonAnywhere" \
    --notaryURL="https://mauricelab.pythonanywhere.
    com/serebo_notary/services/call/xmlrpc" \
    --owner="Maurice HT Ling" \
    --email="mauriceling@acm.org" \
    --bbpath='serebo_blackbox\\blackbox.sdb'

@param owner String: Owner's or administrator's name.
@param email String: Owner's or administrator's email.
@param alias String: Alias for this SEREBO Notary.
@param notaryURL String: URL for SEREBO Notary web
service.Default="https://mauricelab.pythonanywhere.com/
serebo_notary/services/call/xmlrpc"
@param bbpath String: Path to SEREBO black box. Default
= 'serebo_blackbox\\blackbox.sdb'.
'''
db = bb.connectDB(bbpath)
owner = str(owner)
email = str(email)
sqlstmt = "select value from metadata where
    key='blackboxID'"
blackboxID = [row
```

```
                    for row in db.cur.execute(sqlstmt)][0][0]
data = bb.systemData()
architecture = data['architecture']
machine = data['machine']
node = data['node']
platform = data['platform']
processor = data['processor']
try:
    notaryURL, notaryAuthorization, dtstamp) = \
        notary.registerBlackbox(blackboxID, owner, email,
                                architecture, machine,
                                node, platform,
                                processor, notaryURL)
    sqlstmt = '''insert into notary (dtstamp, alias,
        owner, email, notaryDTS, notaryAuthorization,
        notaryURL) values (?,?,?,?,?,?,?)'''
    sqldata = (db.dtStamp(), alias, owner, email,
               dtstamp, notaryAuthorization, notaryURL)
    db.cur.execute(sqlstmt, sqldata)
    db.conn.commit()
    rstring = 'Register SEREBO Black Box with SEREBO
        Notary'
    description = ['Notary URL: %s' % str(notaryURL),
                   'Notary Authorization: %s' % \
                       str(notaryAuthorization),
                   'Notary Date Time Stamp %s' %\
                       str(dtstamp)]
    description = ' | '.join(description)
    rdata = bb.insertFText(db, rstring, description)
    print('')
    print('Registering SEREBO Black Box with SEREBO
        Notary...')
    return {'SEREBO Black Box': db,
            'Black Box Path': str(db.path),
            'Black Box ID': str(blackboxID),
            'Notary URL': str(notaryURL),
            'Notary Authorization': \
                str(notaryAuthorization),
            'Notary Date Time Stamp': str(dtstamp)}
except:
```

```
        print('Registration failed - likely to be SEREBO
            Notary error or XMLRPC error.')
        return {'SEREBO Black Box': db,
                'Black Box Path': str(db.path),
                'Black Box ID': str(blackboxID),
                'Notary URL': str(notaryURL)}

def selfSign(bbpath='serebo_blackbox\\blackbox.sdb'):
    '''
    Function to self-sign (self notarization) SEREBO Black
    Box.

    Usage:

        python serebo.py selfsign \
        --bbpath=<path to SEREBO black box>

    For example:

        python serebo.py selfsign \
        --bbpath='serebo_blackbox\\blackbox.sdb'

    @param bbpath String: Path to SEREBO black box. Default
    = 'serebo_blackbox\\blackbox.sdb'.
    '''
    db = bb.connectDB(bbpath)
    rstring = bb.randomString(db, 32)
    rdata    =    bb.insertFText(db,    rstring,    'Self
    notarization')
    print('')
    print('Self-Signing / Self-Notarization ...')
    return {'SEREBO Black Box': db,
            'Black Box Path': str(db.path),
            'Date Time Stamp': str(rdata['DateTimeStamp']),
            'Random String': str(rstring)}

def notarizeBlackbox(alias,
                bbpath='serebo_blackbox\\blackbox.sdb'):
    '''
```

```
Function to notarize SEREBO Black Box with SEREBO
Notary.

Usage:

    python serebo.py notarizebb \
    --alias=<alias for SEREBO Notary> \
    --bbpath=<path to SEREBO black box>

For example:

    python serebo.py notarizebb \
    --alias="NotaryPythonAnywhere" \
    --bbpath='serebo_blackbox\\blackbox.sdb'

@param alias String: Alias for this SEREBO Notary.
@param bbpath String: Path to SEREBO black box. Default
= 'serebo_blackbox\\blackbox.sdb'.
'''
db = bb.connectDB(bbpath)
sqlstmt = "select value from metadata where
    key='blackboxID'"
blackboxID = [row
            for row in db.cur.execute(sqlstmt)][0][0]
try:
    sqlstmt = "select notaryAuthorization, notaryURL
from notary where alias='%s'" % str(alias)
    sqlresult = [row
                for row in db.cur.execute(sqlstmt)][0]
    notaryAuthorization = sqlresult[0]
    notaryURL = sqlresult[1]
except IndexError:
    print('Notary authorization or Notary URL not found
        for the given alias')
    return {'SEREBO Black Box': db,
            'Black Box Path': str(db.path),
            'Notary Alias': str(alias)}
dtstampBB = bb.dateTime(db)
codeBB = bb.randomString(db, 32)
try:
```

```
            (notaryURL, dtstampNS, codeNS, codeCommon) = \
                notary.notarizeBB(blackboxID,
                                  notaryAuthorization,
                                  dtstampBB, codeBB, notaryURL)
            description = ['Notarization with SEREBO Notary',
                           'Black Box Code: %s' % codeBB,
                           'Black   Box   Date   Time:   %s'   %
                           dtstampBB,
                           'Notary Code: %s' % codeNS,
                           'Notary Date Time: %s' % dtstampNS,
                           'Notary URL: %s' % notaryURL]
            description = ' | '.join(description)
            rdata = bb.insertFText(db, codeCommon, description)
            print('')
            print('Notarizing SEREBO Black Box with SEREBO
                Notary...')
            return {'SEREBO Black Box': db,
                    'Black Box Path': str(db.path),
                    'Notary Alias': str(alias),
                    'Notary URL': str(notaryURL),
                    'Notary Authorization': \
                        str(notaryAuthorization),
                    'Notary Date Time Stamp': str(dtstampNS),
                    'Date Time Stamp': str(dtstampBB),
                    'Black Box Code': str(codeBB),
                    'Notary Code': str(codeNS),
                    'Cross-Signing Code': str(codeCommon)}
        except:
            print('Failed in attempt to notarize SEREBO Black
                Box with SEREBO Notary')
            return {'SEREBO Black Box': db,
                    'Black Box Path': str(db.path),
                    'Notary Alias': str(alias),
                    'Notary URL': str(notaryURL),
                    'Notary Authorization': \
                    str(notaryAuthorization)}

def viewRegistration(bbpath='serebo_blackbox  \\blackbox.
sdb'):
    '''
```

```
    Function to view all SEREBO Notary registration for
    this SEREBO Black Box - This does not insert a record
    into SEREBO Black Box.

    Usage:

        python serebo.py viewreg
        --bbpath=<path to SEREBO black box>

    For example:

        python serebo.py viewreg \
        --bbpath='serebo_blackbox\\blackbox.sdb'

    @param bbpath String: Path to SEREBO black box. Default
    = 'serebo_blackbox\\blackbox.sdb'.
    '''
    db = bb.connectDB(bbpath)
    print('')
    print('Black Box Path: %s' % str(bbpath))
    sqlstmt = '''select dtstamp, alias, owner, email,
    notaryDTS, notaryAuthorization, notaryURL from
    notary'''
    print('')
    print('Notary Registration(s) ...')
    for row in db.cur.execute(sqlstmt):
        print('')
        print('Date Time Stamp: %s' % str(row[0]))
        print('Notary Alias: %s' % str(row[1]))
        print('Owner: %s' % str(row[2]))
        print('Email: %s' % str(row[3]))
        print('Notary Date Time Stamp: %s' % str(row[4]))
        print('Notary Authorization: %s' % str(row[5]))
        print('Notary URL: %s' % str(row[6]))

def changeAlias(alias, newalias,
            bbpath='serebo_blackbox\\blackbox.sdb'):
    '''!
    Function to change alias for a specific SEREBO Notary
    registration.
```

```
    Usage:

        python serebo.py changealias
        --alias=<current alias to be changed>
        --newalias=<new alias to change into>
        --bbpath=<path to SEREBO black box>

    For example:

        python serebo.py changealias \
        --alias="NotaryPythonAnywhere" \
        --newalias="testAlias" \
        --bbpath='serebo_blackbox\\blackbox.sdb'

    @param alias String: Current alias for the SEREBO
    Notary to change.
    @param newalias String: New alias for the SEREBO
    Notary.
    @param bbpath String: Path to SEREBO black box. Default
    = 'serebo_blackbox\\blackbox.sdb'.
    '''
    db = bb.connectDB(bbpath)
    alias = str(alias)
    newalias = str(newalias)
    sqlstmt = '''update notary set alias=? where alias=?'''
    db.cur.execute(sqlstmt, (newalias, alias))
    db.conn.commit()
    message = 'Change notary alias from %s to %s' % \
        alias, newalias)
    rdata = bb.insertFText(db, message, 'NA')
    print('')
    return {'SEREBO Black Box': db,
            'Black Box Path': str(db.path),
            'Alias': alias,
            'New Alias': newalias}

def searchMessage(term, mode='like',
                  bbpath='serebo_blackbox\\blackbox.sdb'):
    '''!
```

```
    Function to search SEREBO Black Box for a message -
    This does not insert a record into SEREBO Black Box.

    Usage:

        python serebo.py searchmsg
        --mode=<search mode>
        --term=<search term>
        --bbpath=<path to SEREBO black box>

    For example:

        python serebo.py searchmsg
        --mode='like'
        --term="Change notary alias%"
        --bbpath='serebo_blackbox\\blackbox.sdb'

    @param term String: Case sensitive search term.
    @param mode String: Mode of search. Allowable modes are
    'like' and 'exact'. If mode is 'like', wildcards such
    as '_' (matches any single character) and '%' (matches
    any number of characters). Default = 'like'.
    @param bbpath String: Path to SEREBO black box. Default
    = 'serebo_blackbox\\blackbox.sdb'.
    '''
    db = bb.connectDB(bbpath)
    mode = str(mode)
    term = str(term)
    result = bb.searchDatalog(db, term, 'data', mode)
    print('')
    print('Search Result (Search by Message) ...')
    print('')
    for row in result:
        print('Date Time Stamp: %s' % str(row[1]))
        print('Message: %s' % str(row[3]))
        print('Description: %s' % str(row[4]))
        print('')
def searchDescription(term, mode='like',
                bbpath='serebo_blackbox\\blackbox.sdb'):
    '''
```

Function to search SEREBO Black Box for a description - This does not insert a record into SEREBO Black Box.

Usage:

    python serebo.py searchdesc
    --mode=<search mode>
    --term=<search term>
    --bbpath=<path to SEREBO black box>

For example:

    python serebo.py searchdesc
    --mode='like'
    --term="%NA%"
    --bbpath='serebo_blackbox\\blackbox.sdb'

@param term String: Case sensitive search term.
@param mode String: Mode of search. Allowable modes are 'like' and 'exact'. If mode is 'like', wildcards such as '_' (matches any single character) and '%' (matches any number of characters). Default = 'like'.
@param bbpath String: Path to SEREBO black box. Default = 'serebo_blackbox\\blackbox.sdb'.
'''
```
db = bb.connectDB(bbpath)
mode = str(mode)
term = str(term)
result = bb.searchDatalog(db, term, 'description', mode)
print('')
print('Search Result (Search by Description) ...')
print('')
for row in result:
    print('Date Time Stamp: %s' % str(row[1]))
    print('Message: %s' % str(row[3]))
    print('Description: %s' % str(row[4]))
    print('')

def searchFile(filepath,
```

```
                bbpath='serebo_blackbox\\blackbox.sdb'):
'''
Function to search SEREBO Black Box for a file logging
event - This does not insert a record into SEREBO Black
Box.

Usage:

    python serebo.py searchfile
    --filepath=<path to file for searching>
    --bbpath=<path to SEREBO black box>

For example:

    python serebo.py searchfile
    --filepath=doxygen_serebo
    --bbpath='serebo_blackbox\\blackbox.sdb'

@param fileapth String: Path of file to search in
SEREBO black box.
@param bbpath String: Path to SEREBO black box. Default
= 'serebo_blackbox\\blackbox.sdb'.
'''
db = bb.connectDB(bbpath)
filepath = str(filepath)
absPath = bb.absolutePath(filepath)
fHash = bb.fileHash(absPath)
result = bb.searchDatalog(db, fHash, 'data', 'exact')
print('')
print('Search Result (Search by File) ...')
print('')
print('File Path: %s' % filepath)
print('Absolute File Path: %s' % absPath)
print('')
for row in result:
    print('Date Time Stamp: %s' % str(row[1]))
    print('Message: %s' % str(row[3]))
    print('Description: %s' % str(row[4]))
    print('')
```

```python
def auditCount(bbpath='serebo_blackbox\\blackbox.sdb'):
    '''!
    Function to check for equal numbers of records in data
    log and blockchain in SEREBO Black Box - should have
    the same number of records. This does not insert a
    record into SEREBO Black Box.

    Usage:

        python serebo.py audit_count
        --bbpath=<path to SEREBO black box>

    For example:

        python serebo.py audit_count \
        --bbpath='serebo_blackbox\\blackbox.sdb'

    @param bbpath String: Path to SEREBO black box. Default
    = 'serebo_blackbox\\blackbox.sdb'.
    '''
    db = bb.connectDB(bbpath)
    sqlstmtA = 'select ID, dtstamp from datalog'
    sqlresultA = {}
    for row in db.cur.execute(sqlstmtA):
        sqlresultA[row[0]] = row[1]
    sqlstmtB = 'select c_ID, c_dtstamp from blockchain'
    sqlresultB = {}
    for row in db.cur.execute(sqlstmtB):
        sqlresultB[row[0]] = row[1]
    print('')
    print('Audit SEREBO Black Box Data Count ...')
    print('')
    if len(sqlresultA) == len(sqlresultB):
        for k in sqlresultA:
            if sqlresultA[k] != sqlresultB[k]:
                print('Date time stamp mismatch')
                print('Datalog record number %s' % str(k))
                print('Datalog date time stamp: %s' % \
                    str(sqlresultA[k]))
                print('Blockchain date time stamp: %s' % \
```

```
                        str(sqlresultB[k])) 
                else: 
                    print('Date time stamp match - Record %s' 
                        % \
                        str(k)) 
        print('Number of records in datalog matches the 
            number of records in blockchain') 
    else: 
        if len(sqlresultA) > len(sqlresultB): 
            print('Number of records in datalog MORE than the 
                number of records in blockchain') 
        elif len(sqlresultA) < len(sqlresultB): 
            print('Number of records in datalog LESS than the 
                number of records in blockchain') 

def auditDatahash(bbpath='serebo_blackbox\\blackbox.sdb'): 
    '''
    Function to check for accuracy of hash generations in 
    data log within SEREBO Black Box - recorded hash in 
    data log and computed hash should be identical. This 
    does not insert a record into SEREBO Black Box. 

    Usage: 

        python serebo.py audit_datahash 
        --bbpath=<path to SEREBO black box> 

    For example: 

        python serebo.py audit_datahash \
        --bbpath='serebo_blackbox\\blackbox.sdb' 

    @param bbpath String: Path to SEREBO black box. Default 
    = 'serebo_blackbox\\blackbox.sdb'. 
    '''
    db = bb.connectDB(bbpath) 
    sqlstmt = '''select ID, dtstamp, data, description, 
        hash from datalog''' 
    print('') 
    print('Audit SEREBO Black Box Data Log Records ...') 
```

```
        print('')
        for row in db.cur.execute(sqlstmt):
            ID = str(row[0])
        dtstamp = str(row[1])
        data = str(row[2])
        description = str(row[3])
        rHash = str(row[4])
        dhash = bytes(dtstamp, 'utf-8') + \
                bytes(data, 'utf-8') + \
                bytes(description, 'utf-8')
        tHash = db.hash(dhash)
        if tHash == rHash:
            print('Verified record %s in data log' % ID)
        else:
            print('ERROR in record %s in data log' % ID)
            print('Hash in record: %s' % rHash)
            print('Computed hash: %s' % tHash)

def dumpHash(outputf,
             bbpath='serebo_blackbox\\blackbox.sdb'):
    '''
    Function to write out record hash from SEREBO Black Box
    into a file - This does not insert a record into SEREBO
    Black Box.

    Usage:

        python serebo.py dumphash
        --outputf=<output file path>
        --bbpath=<path to SEREBO black box>

    For example:

        python serebo.py dumphash \
        --outputf=sereboBB_hash \
        --bbpath='serebo_blackbox\\blackbox.sdb'
    @param outputf String: Output file path. Default =
    sereboBB_hash
    @param bbpath String: Path to SEREBO black box. Default
    = 'serebo_blackbox\\blackbox.sdb'.
```

```
    '''
    db = bb.connectDB(bbpath)
    outputf = str(outputf)
    outputf = bb.absolutePath(outputf)
    outf = open(outputf, 'w')
    sqlstmt = 'select ID, dtstamp, hash from datalog'
    count = 0
    for row in db.cur.execute(sqlstmt):
        data = [str(row[0]), str(row[1]), str(row[2])]
        data = ' | '.join(data)
        outf.write(data + '\n')
        count = count + 1
    outf.close()
    print('')
    print('Dump SEREBO Black Box Data Log Hashes ...')
    print('')
    return {'SEREBO Black Box': db,
            'Black Box Path': str(db.path),
            'Output File Path': outputf,
            'Number of Records': str(count)}

def auditDataBlockchain(bbpath='serebo blackbox\\
                        blackbox.sdb'):
    '''
    Function to check for accuracy in data log and
    blockchain mapping in SEREBO Black Box - recorded hash
    in data log and data in blockchain should be identical.
    This does not insert a record into SEREBO Black Box.

    Usage:

        python serebo.py audit_data_blockchain
        --bbpath=<path to SEREBO black box>

    For example:

        python serebo.py audit_data_blockchain \
        --bbpath='serebo_blackbox\\blackbox.sdb'
```

```python
    @param bbpath String: Path to SEREBO black box. Default
    = 'serebo_blackbox\\blackbox.sdb'.
    '''
    db = bb.connectDB(bbpath)
    sqlstmt = '''select datalog.ID, datalog.dtstamp,
        datalog.hash, blockchain.c_dtstamp, blockchain.data
        from datalog inner join blockchain where
        datalog.ID=blockchain.c_ID and
        datalog.dtstamp=blockchain.c_dtstamp'''
    print('')
    print('Audit SEREBO Black Box - Accuracy in Data Log to
        Blockchain Mapping...')
    print('')
    for row in db.cur.execute(sqlstmt):
        dID = str(row[0])
        ddtstamp = str(row[1])
        dhash = str(row[2])
        bdtstamp = str(row[3])
        bhash = str(row[4])
        if dhash == bhash:
            print('Verified record %s mapping' % dID)
        else:
            print('ERROR in record %s mapping' % dID)
            print('Hash in Data Log: %s' % dHash)
            print('Data in Blockchain: %s' % bHash)

def auditBlockchainHash(bbpath='serebo_blackbox\\
                    blackbox.sdb'):
    '''
    Function to check for accuracy in blockchain hash
    Generation within SEREBO Black Box - recorded hash in
    blockchain and computed hash should be identical. This
    does not insert a record into SEREBO Black Box.

    Usage:

        python serebo.py audit_blockchainhash
        --bbpath=<path to SEREBO black box>

    For example:
```

```
        python serebo.py audit_blockchainhash
        --bbpath='serebo_blackbox\\blackbox.sdb'

    @param bbpath String: Path to SEREBO black box. Default
    = 'serebo_blackbox\\blackbox.sdb'.
    '''
    db = bb.connectDB(bbpath)
    sqlstmt = '''select c_ID, p_dtstamp, p_randomstring,
        p_hash, data, c_hash from blockchain'''
    print('')
    print('Audit SEREBO Black Box Blockchain hashes ...')
    print('')
    for row in db.cur.execute(sqlstmt):
        ID = str(row[0])
        p_dtstamp = str(row[1])
        p_randomstring = str(row[2])
        p_hash = str(row[3])
        data = str(row[4])
        c_hash = str(row[5])
        dhash = ''.join([str(p_dtstamp),str(p_randomstring),
                    str(p_hash), str(data)])
        dhash = bytes(dhash, 'utf-8')
        tHash = db.hash(dhash)
        if tHash == c_hash:
            print('Verified record %s in Blockchain' % ID)
    else:
            print('ERROR in record %s in Blockchain' % ID)
            print('Hash in record: %s' % c_hash)
            print('Computed hash: %s' % tHash)
def checkHash(hashfile,
            bbpath='serebo_blackbox\\blackbox.sdb'):
    '''!
    Function to compare record hash from SEREBO Black Box
    with that in a hash file. This does not insert a record
    into SEREBO Black Box.

    Usage:

        python serebo.py checkhash
        --hashfile=<path to hash file>
```

```
    --bbpath=<path to SEREBO black box>

For example:

    python serebo.py checkhash \
    --hashfile=sereboBB_hash \
    --bbpath='serebo_blackbox\\blackbox.sdb'

@param hashfile String: File path to hash file.
@param bbpath String: Path to SEREBO black box. Default
= 'serebo_blackbox\\blackbox.sdb'.
'''
db = bb.connectDB(bbpath)
hashfile= str(hashfile)
hashfile = bb.absolutePath(hashfile)
print('')
print('Compare record hash from SEREBO Black Box with
    that in a hash file...')
print('')
hf = open(hashfile, 'r')
for record in hf:
    record = [str(d.strip())
             for d in record[:-1].split('|')]
    ID = record[0]
    dtstamp = record[1]
    thash = record[2]
    sqlstmt = """select hash from datalog where ID='%s'
        and dtstamp='%s'""" % (ID, dtstamp)
    dhash  =  [row  for  row  in  db.cur.execute
    sqlstmt)][0][0]
    dhash = str(dhash)
    if thash == dhash:
    print('Verified record %s hash between Data Log
        and Hash file' % ID)
else:
    print('ERROR in record %s' % ID)
    print('Hash in Hash File: %s' % thash)
    print('Hash in Data Log: %s' % dhash)

def auditBlockchainFlow(bbpath='serebo_blackbox\\
```

```
                blackbox.sdb'):
'''!
Function to trace the decendancy of blockchain records
(also known as blocks) within SEREBO Black Box -
decandency from first block should be traceable to the
last / latest block. This does not insert a record into
SEREBO Black Box.

Usage:

    python serebo.py audit_blockchainflow
    --bbpath=<path to SEREBO black box>

For example:

    python serebo.py audit_blockchainflow
    --bbpath='serebo_blackbox\\blackbox.sdb'

@param bbpath String: Path to SEREBO black box. Default
= 'serebo_blackbox\\blackbox.sdb'.
'''
db = bb.connectDB(bbpath)
sqlstmt = '''select max(c_ID) from blockchain'''
print('')
print("Trace SEREBO Black Box Blockchain's block
    decendancy ...")
print('')
maxID = [row for row in db.cur.execute(sqlstmt)][0][0]
maxID = int(maxID)
for i in range(1, maxID, 1):
    # Get parent data from parent block
    sqlstmt = """select c_ID, c_dtstamp, c_randomstring,
        c_hash from blockchain where c_ID=%s""" % str(i)
    #print(sqlstmt)
    p_data = [row for row in db.cur.execute(sqlstmt)][0]
    pc_ID = str(p_data[0])
    pc_dtstamp = str(p_data[1])
    pc_randomstring = str(p_data[2])
    pc_hash = str(p_data[3])
    # Get parent data from current / child block
```

```python
        sqlstmt = """select p_ID, p_dtstamp, p_randomstring,
            p_hash from blockchain where c_ID=%s""" % \
            str(i+1)
        #print(sqlstmt)
        c_data = [row for row in db.cur.execute(sqlstmt)][0]
        p_ID = str(c_data[0])
        p_dtstamp = str(c_data[1])
        p_randomstring = str(c_data[2])
        p_hash = str(c_data[3])
        # Compare parental block record and parent data in
        # current record
        if (p_ID == pc_ID) and \
            (p_dtstamp == pc_dtstamp) and \
            (p_randomstring == pc_randomstring) and \
            (p_hash == pc_hash):
            print('Verified - Record %s was used as parent
                record in record %s' % (str(i), str(i+1)))
        else:
            print('ERROR in record %s' % str(i+1))
            print('Parent ID in record %s: %s' % \
                (str(i+1), str(i)))
            print('Parent date time stamp in record %s:
            %s' \
                % (str(i+1), p_dtstamp))
            print('Actual date time stamp in record %s:
            %s' \
                % (str(i), pc_dtstamp))
            print('Parent random string in record %s:
            %s' % \
                (str(i+1), p_randomstring))
            print('Actual random string in record %s:
            %s'% \
                (str(i), pc_randomstring))
            print('Parent hash in record %s: %s' % \
                (str(i+1), p_hash))

            print('Actual hash in record %s: %s' % \
                (str(i), pc_hash))

def NTPSign(bbpath='serebo_blackbox\\blackbox.sdb'):
```

```
'''!
Function to self-sign (self notarization) SEREBO Black
Box using NTP (Network Time Protocol) server.

Usage:

    python serebo.py ntpsign
    --bbpath=<path to SEREBO black box>

For example:

    python serebo.py ntpsign
    --bbpath='serebo_blackbox\\blackbox.sdb'

@param bbpath String: Path to SEREBO black box. Default
= 'serebo_blackbox\\blackbox.sdb'.
'''
db = bb.connectDB(bbpath)
ntp = bb.ntplib.NTPClient()
rstring = bb.randomString(db, 32)
response = ntp.request('pool.ntp.org', version=3)
dtstamp = bb.gmtime(response.tx_time)
ntp_ip = bb.ntplib.ref_id_to_text(response.ref_id)
description = ['NTP server (self) notarization',
               'Seconds Since Epoch: %s' % \
                str(response.tx_time),
               'NTP Date Time: %s' % str(dtstamp),
               'NTP Server IP: %s' % str(ntp_ip)]
description = ' | '.join(description)
rdata = bb.insertFText(db, rstring, description)
print('')
print('Self-Signing / Self-Notarization ...')
print('')

return {'SEREBO Black Box': db,
        'Black Box Path': str(db.path),
        'Date Time Stamp': str(rdata['DateTimeStamp']),
        'Random String': str(rstring),
        'Seconds Since Epoch': str(response.tx_time),
        'NTP Date Time': str(dtstamp),
```

```
                'NTP Server IP': str(ntp_ip)}

def backup(backuppath='blackbox_backup.sdb',
           bbpath='serebo_blackbox\\blackbox.sdb'):
    '''
    Function to backup SEREBO Black Box - This does not
    insert A record into SEREBO Black Box.

    Usage:

        python serebo.py backup
        --backuppath=<path for backed-up SEREBO black box>
        --bbpath=<path to SEREBO black box>

    For example:

        python serebo.py backup
        --backuppath='blackbox_backup.sdb'
        --bbpath='serebo_blackbox\\blackbox.sdb'

    @param backuppath String: Path for backed-up SEREBO
    black box. Default = 'blackbox_backup.sdb'
    @param bbpath String: Path to SEREBO black box. Default
    = 'serebo_blackbox\\blackbox.sdb'.
    '''
    print('')
    print('Backup SEREBO Black Box ...')
    print('')
    if backuppath != bbpath:
        (bbpath, backuppath) = bb.backup(bbpath, backuppath)
        return {'Black Box Path': bbpath,
                'Backup Path': backuppath}
    else:
        print('Backup path cannot be the same as SEREBO
            Black Box path')
        bbpath = bb.absolutePath(bbpath)
        backuppath = bb.absolutePath(backuppath)
        return {'Black Box Path': bbpath,
                'Backup Path': backuppath}
```

```
def dump(dumpfolder='.', fileprefix='dumpBB',
         bbpath='serebo_blackbox\\blackbox.sdb'):
    '''
    Function to dump individual data tables from SEREBO
    Black Box into text files - This does not insert a
    record into SEREBO Black Box.

    Usage:

        python serebo.py dump
        --dumpfolder=<folder to save dump files>
        --fileprefix=<prefix for individual dump files>
        --bbpath=<path to SEREBO black box>

    For example:

        python serebo.py dump
        --dumpfolder='.'
        --fileprefix='dumpBB'
        --bbpath='serebo_blackbox\\blackbox.sdb'

    @param dumpfolder String: Folder to save dump files.
    Default = '.' (current working directory).
    @param fileprefix String: Prefix for individual dump
    files. Default = 'dumpBB'.
    @param bbpath String: Path to SEREBO black box. Default
    = 'serebo_blackbox\\blackbox.sdb'.
    '''
    db = bb.connectDB(bbpath)
    tableSet = {'metadata': ['key', 'value'],
                'notary': ['dtstamp',
                'alias',
                'owner',
                'email',
                'notaryDTS',
                'notaryAuthorization',
                'notaryURL'],
        'systemdata': ['dtstamp', 'key', 'value'],
        'datalog': ['dtstamp',
                'hash',
```

```
                    'data',
                    'description'],
        'blockchain': ['c_ID', 'c_dtstamp',
                       'c_randomstring', 'c_hash',
                       'p_ID', 'p_dtstamp',
                       'p_randomstring', 'p_hash',
                       'data'],
        'eventlog': ['dtstamp', 'fID',
                     'description'],
        'eventlog_datamap': ['dtstamp', 'fID',
                             'key', 'value']}
    print('')
    print('Dump out data (text backup) from SEREBO Black
        Box ...')
    print('')
    for tableName in tableSet:
        outputfile = [dumpfolder,
                      fileprefix + '_' + tableName + '.csv']
        outputfile = os.sep.join(outputfile)
        (outputfile, count) = bb.dumpTable(db, tableName,
                      tableSet[tableName], outputfile)
        print('%s table dumped into %s' % \
              (tableName, outputfile))
        print('Number of records dumped: %s' % count)
        print('')

def auditRegister(alias,
                  bbpath='serebo_blackbox\\blackbox.sdb'):
    '''
    Function to check for SEREBO Black Box registration
    with SEREBO Notary - This does not insert a record into
    SEREBO Black Box.

    Usage:

        python serebo.py audit_register \
        --alias=<alias for SEREBO Notary> \
        --bbpath=<path to SEREBO black box>

    For example:
```

```
        python serebo.py audit_register \
        --alias="NotaryPythonAnywhere" \
        --bbpath='serebo_blackbox\\blackbox.sdb'

    @param alias String: Alias for this SEREBO Notary.
    @param bbpath String: Path to SEREBO black box. Default
    = 'serebo_blackbox\\blackbox.sdb'.
    '''
    db = bb.connectDB(bbpath)
    sqlstmt = "select value from metadata where
        key='blackboxID'"
    blackboxID = [row
                  for row in db.cur.execute(sqlstmt)][0][0]
    try:
        sqlstmt = "select  notaryAuthorization,  notaryURL
from notary where alias='%s'" % str(alias)
        sqlresult = [row
                     for row in db.cur.execute(sqlstmt)][0]
        notaryAuthorization = sqlresult[0]
        notaryURL = sqlresult[1]
    except IndexError:
        print('Notary authorization or Notary URL not found
              for the given alias')
        return {'SEREBO Black Box': db,
                'Black Box Path': str(db.path),
                'Notary Alias': str(alias)}
    try:
        presence = notary.checkRegistration(blackboxID,
                   notaryAuthorization, notaryURL)
        if presence:
           message = 'Registration found in SEREBO Notary'
        else:
           message = 'Registration NOT found in SEREBO
               Notary'
        print('')

        print('Checking  SEREBO  Black  Box  registration  in
            SEREBO Notary...')
        return {'SEREBO Black Box': db,
                'Black Box Path': str(db.path),
```

```
                    'Notary Alias': str(alias),
                    'Notary URL': str(notaryURL),
                    'Notary Authorization': \
                        str(notaryAuthorization),
                    'Status': message}
    except:
        print('Failed in checking SEREBO Black Box
            registration in SEREBO Notary')
        return {'SEREBO Black Box': db,
                'Black Box Path': str(db.path),
                'Notary Alias': str(alias),
                'Notary URL': str(notaryURL),
                'Notary Authorization': \
                    str(notaryAuthorization)}

def viewSelfNotarizations(bbpath='serebo_blackbox\\
                          blackbox.sdb'):
    '''!
    Function to view all self notarizations for this SEREBO
    Black Box - This does not insert a record into SEREBO
    Black Box.

    Usage:

        python serebo.py viewselfnote
        --bbpath=<path to SEREBO black box>

    For example:

        python serebo.py viewselfnote
        --bbpath='serebo_blackbox\\blackbox.sdb'

    @param bbpath String: Path to SEREBO black box. Default
    = 'serebo_blackbox\\blackbox.sdb'.
    '''
    db = bb.connectDB(bbpath)
    print('')
    print('Black Box Path: %s' % str(bbpath))
    sqlstmt = """select dtstamp, data from datalog where
        description like 'Self notarization'"""
```

```
    print('')
    print('Self Notarization(s) ...')
    for row in db.cur.execute(sqlstmt):
        print('')
        print('Date Time Stamp: %s' % str(row[0]))
        print('Hash: %s' % str(row[1]))

def viewNTPNotarizations(bbpath='serebo_blackbox\\
                        blackbox.sdb'):
    '''
    Function to view all self-notarization(s) by NTP time
    server for this SEREBO Black Box - This does not insert
    a record into SEREBO Black Box.

    Usage:

        python serebo.py viewntpnote
        --bbpath=<path to SEREBO black box>

    For example:

        python serebo.py viewntpnote
        --bbpath='serebo_blackbox\\blackbox.sdb'

    @param bbpath String: Path to SEREBO black box. Default
    = 'serebo_blackbox\\blackbox.sdb'.
    '''
    db = bb.connectDB(bbpath)
    print('')
    print('Black Box Path: %s' % str(bbpath))
    sqlstmt = """select dtstamp, data, description from
        datalog where description like 'NTP server (self)
        notarization%'"""
    print('')
    print('Self-Notarization(s) by NTP Time Server(s) ...')
    for row in db.cur.execute(sqlstmt):
        description = [x.strip()
                       for x in str(row[2]).split('|')]
        print('')
        print('Date Time Stamp: %s' % str(row[0]))
```

```python
    print('Random Code: %s' % str(row[1]))
    print('NTP Seconds Since Epoch: %s' % description[1])
    print('NTP Date Time: %s' % description[2])
    print('NTP Server IP: %s' % description[3])

def viewNotaryNotarizations(bbpath='serebo_blackbox\\
                            blackbox.sdb'):
    '''!
    Function to view all notarizations by SEREBO Notary for
    this SEREBO Black Box - This does not insert a record
    into SEREBO Black Box.

    Usage:

        python serebo.py viewsnnote \
        --bbpath=<path to SEREBO black box>

    For example:

        python serebo.py viewsnnote \
        --bbpath='serebo_blackbox\\blackbox.sdb'

    @param bbpath String: Path to SEREBO black box. Default
    = 'serebo_blackbox\\blackbox.sdb'.
    '''
    db = bb.connectDB(bbpath)
    print('')
    print('Black Box Path: %s' % str(bbpath))
    sqlstmt = """select dtstamp, data, description from
        datalog where description like 'Notarization with
        SEREBO Notary%'"""
    print('')
    print('Notarization(s) by SEREBO Notary(ies) ...')
    for row in db.cur.execute(sqlstmt):
        description = [x.strip()
                       for x in str(row[2]).split('|')]
        print('')
        print('Date Time Stamp: %s' % str(row[0]))
        print('Common Code: %s' % str(row[1]))
        print(description[1]) # Black Box Code
```

```
        print(description[2])  # Black Box Date Time
        print(description[3])  # Notary Code
        print(description[4])  # Notary Date Time
        print(description[5])  # Notary URL

def _auditSingleNotarizeBB(blackboxID, notaryAuthorization,
        notaryURL, BBCode, NCode, CommonCode):
    '''!
    Private function - communicate with SEREBO Notary to
    check for SEREBO Black Box notarization record.

    @param blackboxID String: ID of SEREBO black box -
    found in metadata table in SEREBO black box database.
    @param notaryAuthorization String: Notary authorization
    code of SEREBO black box (generated during black box
    registration - found in metadata table in SEREBO black
    box database.
    @param notaryURL String: URL for SEREBO Notary web
    service.
    @param BBCode String: Notarization code from SEREBO
    Black Box.
    @param NCode String: Notarization code from SEREBO
    Notary.
    @param CommonCode String: Cross-Signing code from
    SEREBO Notary.
    @returns: 'True' if SEREBO Black Box notarization is
    found in SEREBO Notary. 'False' if SEREBO Black Box
    notarization is not found in SEREBO Notary. 'Failed' if
    there is any errors, such as network error.
    '''
    try:
        presence = notary.checkNotarization(blackboxID,
                            notaryAuthorization,
                                BBCode,
                                NCode,
                                CommonCode,
                                notaryURL)
        return presence
    except:
        return 'Failed'
```

```python
def
auditNotarizeBB(bbpath='serebo_blackbox\\blackbox.sdb'):
    '''!
    Function to view all notarizations by SEREBO Notary for
    this SEREBO Black Box - This does not insert a record
    into SEREBO Black Box.

    Usage:

        python serebo.py audit_notarizebb
        --bbpath=<path to SEREBO black box>

    For example:

        python serebo.py audit_notarizebb
        --bbpath='serebo_blackbox\\blackbox.sdb'

    @param bbpath String: Path to SEREBO black box. Default
    = 'serebo_blackbox\\blackbox.sdb'.
    '''
    db = bb.connectDB(bbpath)
    sqlstmt = "select value from metadata where
        key='blackboxID'"
    blackboxID = [row
                  for row in db.cur.execute(sqlstmt)][0][0]
    print('')
    print('Black Box Path: %s' % str(bbpath))
    sqlstmtA = """select dtstamp, data, description from
        datalog where description like 'Notarization with
        SEREBO Notary%'"""
    dataA = [row for row in db.cur.execute(sqlstmtA)]
    print('')
    print('Notarization(s) by SEREBO Notary(ies) ...')
    for row in dataA:
        description = [x.strip()
                       for x in str(row[2]).split('|')]
        try:
            notaryURL = description[5].split(': ')[1].strip()
            sqlstmt = "select notaryAuthorization from notary
                where notaryURL='%s'" % str(notaryURL)
```

```python
            sqlresult = [row 
                        for row in db.cur.execute(sqlstmt)][0]
            notaryAuthorization = sqlresult[0]
        except IndexError:
            print('Notary authorization not found for the given 
                Notary URL')
            return {'SEREBO Black Box': db, 
                    'Black Box Path': str(db.path),
                    'Notary URL': str(notaryURL)}
        presence = _auditSingleNotarizeBB(blackboxID, 
                    notaryAuthorization, notaryURL,
                    description[1].split(': ')[1].strip(),
                    description[3].split(': ')[1].strip(),
                    str(row[1]))
        if presence == 'True':
            message = 'Notarization record is found in SEREBO 
                Notary'
        elif presence == 'False':
            message = 'Notarization record is NOT found in 
                SEREBO Notary'
        elif presence == 'Failed':
            message = 'Unspecified error - does not mean that 
                notarization record is not found. It may mean 
                network error.'
        print('')
        print('Date Time Stamp: %s' % str(row[0]))
        print('Common Code: %s' % str(row[1]))
        print(description[1]) # Black Box Code
        print(description[2]) # Black Box Date Time
        print(description[3]) # Notary Code
        print(description[4]) # Notary Date Time
        print(description[5]) # Notary URL
        print('Status: %s' % message)
if __name__ == '__main__':
    exposed_functions = {\
        'audit_blockchainflow': auditBlockchainFlow,
        'audit_blockchainhash': auditBlockchainHash,
        'audit_count': auditCount,
        'audit_data_blockchain': auditDataBlockchain,
        'audit_datahash': auditDatahash,
```

```
    'audit_notarizebb': auditNotarizeBB,
    'audit_register': auditRegister,
    'backup': backup,
    'changealias': changeAlias,
    'checkhash': checkHash,
    'dump': dump,
    'dumphash': dumpHash,
    'fhash': fileHash,
    'init': initialize,
    'intext': insertText,
    'localcode': localCode,
    'localdts': localDTS,
    'logfile': logFile,
    'notarizebb': notarizeBlackbox,
    'ntpsign': NTPSign,
    'register': registerBlackbox,
    'searchmsg': searchMessage,
    'searchdesc': searchDescription,
    'searchfile': searchFile,
    'selfsign': selfSign,
    'shash': stringHash,
    'sysdata': systemData,
    'sysrecord': systemRecord,
    'viewntpnote': viewNTPNotarizations,
    'viewselfnote': viewSelfNotarizations,
    'viewsnnote': viewNotaryNotarizations,
    'viewreg': viewRegistration}
fire.Fire(exposed_functions)
```

File Name: SEREBO_NOTARY_API.PY

```
'''!
Secured Recorder Box (SEREBO) Notary Communicator

Date created: 19th May 2018

License: GNU General Public License version 3 for
academic or not-for-profit use only
```

```
SEREBO is free software: you can redistribute it and/or
modify it under the terms of the GNU General Public
License as published by the Free Software Foundation,
either version 3 of the License, or (at your option) any
later version.

This program is distributed in the hope that it will be
useful, but WITHOUT ANY WARRANTY; without even the implied
warranty of MERCHANTABILITY or FITNESS FOR A PARTICULAR
PURPOSE. See the GNU General Public License for more
details.

You should have received a copy of the GNU General
Public License along with this program. If not, see
<http://www.gnu.org/licenses/>.
'''

from xmlrpc.client import import ServerProxy

def registerBlackbox(blackboxID, owner, email,
                     architecture, machine, node,
                     platform, processor,
                     notaryURL='https://mauricelab.
                     pythonanywhere.com/serebo_notary/
                     services/call/xmlrpc'):
'''!
Function to communicate with SEREBO Notary to register
SEREBO Black Box with SEREBO Notary.

@param blackboxID String: ID of SEREBO black box -
found in metadata table in SEREBO black box database.
@param owner String: Owner's or administrator's name.
@param email String: Owner's or administrator's email.
@param architecture String: Architecture of this
machine - from platform library in Python Standard
Library.
@param machine String: This machine description - from
Platform library in Python Standard Library.
@param node String: This machine's node description -
from platform library in Python Standard Library.
```

```
    @param platform String: This platform description -
    from platform library in Python Standard Library.
    @param processor String: Machine's processor
    description - from platform library in Python Standard
    Library
    @param notaryURL String: URL for SEREBO Notary web
    service.
    Default="https://mauricelab.pythonanywhere.com/
    serebo_notary/services/call/xmlrpc"
    @returns: (URL of SEREBO Notary, Notary authorization
    code, Date time stamp from SEREBO Notary)
    '''
    serv = ServerProxy(notaryURL)
    (notaryAuthorization, dtstamp) = \
        serv.register_blackbox(blackboxID, owner, email,
                               architecture, machine, node,
                               platform, processor)
    return (notaryURL,
            str(notaryAuthorization),
            str(dtstamp))

def notarizeBB(blackboxID, notaryAuthorization, dtstampBB,
        codeBB, notaryURL='https://mauricelab.
               pythonanywhere.com/serebo_notary/
               services/call/xmlrpc'):
    '''!
    Function to communicate with SEREBO Notary to notarize
    SEREBO Black Box with SEREBO Notary.
    @param blackboxID String: ID of SEREBO black box -
    found in metadata table in SEREBO black box database.
    @param notaryAuthorization String: Notary authorization
    code of SEREBO black box (generated during black box
    registration - found in metadata table in SEREBO black
    box database.
    @param dtstampBB String: Date time stamp from SEREBO
    black box.
    @param codeBB String: Notarization code from SEREBO
    black box.
    @param notaryURL String: URL for SEREBO Notary web
```

```
    service.
    Default="https://mauricelab.pythonanywhere.com/
    serebo_notary/services/call/xmlrpc"
    @returns: (URL of SEREBO Notary, Date time stamp from
    SEREBO Notary, Notarization code from SEREBO Notary,
    Cross-Signing code from SEREBO Notary)
    '''
    serv = ServerProxy(notaryURL)
    (dtstampNS, codeNS, codeCommon) = \
        serv.notarizeSereboBB(blackboxID,
        notaryAuthorization, dtstampBB, codeBB)
    return (notaryURL, str(dtstampNS),
            str(codeNS), str(codeCommon))

def checkRegistration(blackboxID, notaryAuthorization,
    notaryURL='https://mauricelab.pythonanywhere.com/
    serebo_notary/services/call/xmlrpc'):
    '''!
    Function to communicate with SEREBO Notary to check for
    SEREBO Black Box registration record.

    @param blackboxID String: ID of SEREBO black box -
    found in metadata table in SEREBO black box database.
    @param notaryAuthorization String: Notary authorization
    code of SEREBO black box (generated during black box
    registration - found in metadata table in SEREBO black
    box database.
    @param notaryURL String: URL for SEREBO Notary web
    service.
    Default="https://mauricelab.pythonanywhere.com/
    serebo_notary/services/call/xmlrpc"
    @returns: Boolean flag - True if SEREBO Black Box
    registration is found in SEREBO Notary. False if SEREBO
    Black Box registration is not found in SEREBO Notary.
    '''
    serv = ServerProxy(notaryURL)
    value = serv.checkBlackBoxRegistration(blackboxID,
            notaryAuthorization)
    if value or value == 'True':
        return True
```

```
    elif not value or value == 'False':
        return False

def checkNotarization(blackboxID, notaryAuthorization,
    BBCode, NCode, CommonCode,
    notaryURL='https://mauricelab.pythonanywhere.com/
    serebo_notary/services/call/xmlrpc'):
    '''
    Function to communicate with SEREBO Notary to check for
    SEREBO Black Box notarization record.

    @param blackboxID String: ID of SEREBO black box -
    found in metadata table in SEREBO black box database.
    @param notaryAuthorization String: Notary authorization
    code of SEREBO black box (generated during black box
    registration - found in metadata table in SEREBO black
    box database.
    @param BBCode String: Notarization code from SEREBO
    Black Box.
    @param NCode String: Notarization code from SEREBO
    Notary.
    @param CommonCode String: Cross-Signing code from
    SEREBO Notary.
    @param notaryURL String: URL for SEREBO Notary web
    service.
    Default="https://mauricelab.pythonanywhere.com/
    serebo_notary/services/call/xmlrpc"
    @returns: Boolean flag - True if SEREBO Black Box
    notarization is found in SEREBO Notary. False if SEREBO
    Black Box notarization is not found in SEREBO Notary.
    '''
    serv = ServerProxy(notaryURL)
    value = serv.checkNotarizeSereboBB(blackboxID,
                                      notaryAuthorization,
                                      BBCode, NCode,
                                      CommonCode)
    if value or value == 'True':
        return 'True'
    elif not value or value == 'False':
        return 'False'
```

Code Files for SEREBO BlackBox

File Name: __INIT__.PY

```
'''!
Secured Recorder Box (SEREBO) Black Box

Date created: 17th May 2018

License: GNU General Public License version 3 for
academic or not-for-profit use only

SEREBO is free software: you can redistribute it and/or
modify it under the terms of the GNU General Public
License as published by the Free Software Foundation,
either version 3 of the License, or (at your option) any
later version.

This program is distributed in the hope that it will be
useful, but WITHOUT ANY WARRANTY; without even the implied
warranty of MERCHANTABILITY or FITNESS FOR A PARTICULAR
PURPOSE. See the GNU General Public License for more
details.
    You should have received a copy of the GNU General
Public License along with this program. If not, see
<http://www.gnu.org/licenses/>.
'''

from datetime import datetime
# Metadata
__version__ = '1.0'
__author__ = 'Maurice H.T. Ling <mauriceling@acm.org>'
__maintainer__ = 'Maurice H.T. Ling <mauriceling@acm.org>'
__email__ = 'mauriceling@acm.org'
__copyright__ = '(c) 2018-%s, Maurice H.T. Ling.' % \
(datetime.now().year)
__description__ = '''
```

SEREBO (SEcured REcorder BOx) Black Box is inspired by the black boxes (cockpit voice recorder and flight data recorder) in airliners. The intended purpose is to track and audit research records under the following premise - Given a set of data files, is there a system to log and verify that these files had not been changed or edited since its supposed creation?

SEREBO Black Box aims to address this issue using several approaches. Firstly, the data files can be used to generate a file hash. It is very likely that an edit in the file will result in a different hash. Hence, if a file generates the same hash across two different points in time, it can be safely assumed that the file had not been edited during this time span. Secondly, the file hash has to be securely recorded with amendment protected. SEREBO records the hash and registers the hash into a blockchain. The main concept of blockchain is that the hash of previous (parent) block is concatenated with the data (file hash in this case) of the current block to generate a hash for the current block. Hence, as the blockchain grows, any amendments in earlier blocks can be easily detected - only amendments to the latest block cannot be detected. Therefore, the value of SEREBO lies in its use.'''

```
from . import ntplib
from . import serebo_api
from .serebo_api import absolutePath
from .serebo_api import backup
from .serebo_api import connectDB
from .serebo_api import dateTime
from .serebo_api import dumpTable
from .serebo_api import fileHash
from .serebo_api import gmtime
from .serebo_api import insertFText
from .serebo_api import insertText

from .serebo_api import logFile
from .serebo_api import randomString
from .serebo_api import searchDatalog
```

```
from .serebo_api import stringHash
from .serebo_api import systemData
```

File Name: SEREBO_API.PY

```
'''
Secured Recorder Box (SEREBO) Application Programming
Interface (API)

Date created: 17th May 2018

License: GNU General Public License version 3 for
academic or not-for-profit use only

SEREBO is free software: you can redistribute it and/or
modify it under the terms of the GNU General Public
License as published by the Free Software Foundation,
either version 3 of the License, or (at your option) any
later version.
This program is distributed in the hope that it will be
useful, but WITHOUT ANY WARRANTY; without even the implied
warranty of MERCHANTABILITY or FITNESS FOR A PARTICULAR
PURPOSE. See the GNU General Public License for more
details.
You should have received a copy of the GNU General
Public License along with this program. If not, see
<http://www.gnu.org/licenses/>.
'''
import hashlib
import random
import secrets
import os.path
import time

from . import sereboDB
from .sereboDB import SereboDB

def connectDB(bbpath='serebo_blackbox\\blackbox.sdb'):
```

```
    '''
    Function to connect to SEREBO database - the recorder
    box.

    @param bbpath String: Path to SEREBO black box. Default
    = 'serebo_blackbox\\blackbox.sdb'.
    @return: SEREBO database object
    '''
    bbpath = os.path.abspath(bbpath)
    db = SereboDB(bbpath)
    return db

def systemData():
    '''
    Function to extract data and test hashes of current
    platform.

    @return: Dictionary of system data and test hashs
    '''
    import platform
    data = {\
        'architecture': ':'.join(platform.architecture()),
        'machine': platform.machine(),
        'node': platform.node(),
        'platform': platform.platform(),
        'processor': platform.processor(),
        'python_build': ' > '.join(platform.python_build()),
        'python_compiler': platform.python_compiler(),
        'python_implementation': \
            platform.python_implementation(),
        'python_branch': platform.python_branch(),
        'python_revision': platform.python_revision(),
        'python_version': platform.python_version(),
        'release': platform.release(),
        'system': platform.system(),
        'version': platform.version()
        }
    data['hashdata'] = \
        bytes('''yd6jwAYeqHmzSyxkNOVXTGtDr8dgZIE9LoL9jxRUbq
OEuODCysfeJkLJHy3LuQX3Rp4f1Ms5HcfTDAyjdLSpNVJx2vbks
```

```
      BKAAi5VVkhW7MJ9CtlfZBlBvCYbX8Qk8Jw27fsglmaPmbR9BZQo
      FpuSQxCDF77dmCcbqw5WiKfuTQiUl9PeyHemnMVtsRGKfN2c0x0
      BA54HjOyN30Dy86fJhitrhLsW3wIY9PtzFEcXd1rq36cFKfrNp7
      lRjJzDJ4W8ZCuQY6P3HUM8Eu4fsGytH9WlVmJ1aJGiyPVf1ZAa4
      2yKUnfBUwhFNU1aEtplVeHrQqQvO7tLxyE5Oc8TjRF7sAzQozjV
      bNyhVlxOmhI45pX4qtBA9y9XrHfYJP9RJaprTsnR24g1pOjxVyp
      zEjSGVEh7EKYWXk7fLllwWRkAb7rG5HSEH5gmcsvbpTNNEsXfcm
      myrvvh6i7cfQGPap2XmxjO6VRZg1hkf7yUarltZ1kTdD3pMJRBo
      PpPijuqB1uA''', 'utf-8')
  data['hash_md5'] = \
      hashlib.md5(data['hashdata']).hexdigest()
  data['hash_sha1'] = \
      hashlib.sha1(data['hashdata']).hexdigest()
  data['hash_sha224'] = \
      hashlib.sha224(data['hashdata']).hexdigest()
  data['hash_sha3_224'] = \
      hashlib.sha3_224(data['hashdata']).hexdigest()
  data['hash_sha256'] = \
      hashlib.sha256(data['hashdata']).hexdigest()
  data['hash_sha3_256'] = \
      hashlib.sha3_256(data['hashdata']).hexdigest()
  data['hash_sha384'] = \
      hashlib.sha384(data['hashdata']).hexdigest()
  data['hash_sha3_384'] = \
      hashlib.sha3_384(data['hashdata']).hexdigest()
  data['hash_sha512'] = \
      hashlib.sha512(data['hashdata']).hexdigest()
  data['hash_sha3_512'] = \
      hashlib.sha3_512(data['hashdata']).hexdigest()
  data['hash_blake2b'] = \
      hashlib.blake2b(data['hashdata']).hexdigest()
  data['hash_blake2s'] = \
      hashlib.blake2s(data['hashdata']).hexdigest()
  return data

def insertText(sdb_object, text, description='NA'):
    '''
    Function to insert text string into SEREBO database,
    with 10-random character string suffixing the
    description.
```

A dictionary of items generated will be returned with the following keys: (1) DateTimeStamp is the UTC date time stamp of this event, (2) Data is the given data string to be inserted, (3) UserDescription is the user given explanation string for this event suffixed with a 10-character random string, (4) DataHash is the hash string of Data, (5) ParentBlockID is the ID of the parent block in blockchain, (6) ParentDateTimeStamp is the UTC date time stamp of the parent block in blockchain (which is also the parent insertion event), (7) ParentRandomString is the random string generated in parent block in blockchain, (8) ParentHash is the hash of parent block in blockchain, (9) BlockRandomString is the random string generated for current insertion event, and (10) BlockHash is the block hash of current insertion event in blockchain.

@param sdb_object Object: SEREBO database object.
@param text String: Text string to be inserted.
@param description String: Explanation string for this Entry event. Default = NA.
@return: Dictionary of data generated from this event.
'''
rdata = sdb_object.insertData(text, description, 'text')
return rdata

def insertFText(sdb_object, text, description='NA'):
    '''
    Function to insert text string into SEREBO database, without 10-random character string suffixing the description.

    A dictionary of items generated will be returned with the following keys: (1) DateTimeStamp is the UTC date time stamp of this event, (2) Data is the given data string to be inserted, (3) UserDescription is the user given explanation string for this event, (4) DataHash is the hash string of Data, (5) ParentBlockID is the ID of the parent block in blockchain, (6)

ParentDateTimeStamp is the UTC date time stamp of the parent block in blockchain (which is also the parent insertion event), (7) ParentRandomString is the random string generated in parent block in blockchain, (8) ParentHash is the hash of parent block in blockchain, (9) BlockRandomString is the random string generated for current insertion event, and (10) BlockHash is the block hash of current insertion event in blockchain.

@param sdb_object Object: SEREBO database object.
@param text String: Text string to be inserted.
@param description String: Explanation string for this entry event. Default = NA.
@return: Dictionary of data generated from this event.
'''
```
    rdata = sdb_object.insertData(text, description, 'ftext')
    return rdata


def absolutePath(filepath):
    '''
    Function to convert file path (absolute or relative file path) into absolute file path.
    @param filepath String: File path to be converted.
    @return: Absolute file path.
    '''
    return os.path.abspath(filepath)


def fileHash(filepath):
    '''
    Function to generate a series of 12 hashes for a given file, in the format of <MD5>:<SHA1>:<SHA224>:<SHA3 244>:<SHA256>:<SHA3 256>:<SHA384>:<SHA3 384>:<SHA512>: <SHA3 215>:<Blake 2b>:<Blake 2s>.

    @param filepath String: Path of file for hash generation.
    @return: Hash
    '''
    absPath = absolutePath(filepath)
```

```
md5 = hashlib.md5()
sha1 = hashlib.sha1()
sha224 = hashlib.sha224()
sha3_224 = hashlib.sha3_224()
sha256 = hashlib.sha256()
sha3_256 = hashlib.sha3_256()
sha384 = hashlib.sha384()
sha3_384 = hashlib.sha3_384()
sha512 = hashlib.sha512()
sha3_512 = hashlib.sha3_512()
blake2b = hashlib.blake2b()
blake2s = hashlib.blake2s()
with open(absPath, 'rb') as f:
    while True:
        data = f.read(65536)
        if not data:
           break
        md5.update(data)
        sha1.update(data)
        sha224.update(data)
        sha3_224.update(data)
        sha256.update(data)
        sha3_256.update(data)
        sha384.update(data)
        sha3_384.update(data)
        sha512.update(data)
        sha3_512.update(data)
        blake2b.update(data)
        blake2s.update(data)
x = [md5.hexdigest(),
    sha1.hexdigest(),
    sha224.hexdigest(),
    sha3_224.hexdigest(),
    sha256.hexdigest(),
    sha3_256.hexdigest(),
    sha384.hexdigest(),
    sha3_384.hexdigest(),
    sha512.hexdigest(),
    sha3_512.hexdigest(),
    blake2b.hexdigest(),
```

```
        blake2s.hexdigest()]
    return ':'.join(x)


def logFile(sdb_object, filepath, description='NA'):
    '''!
    Function to logging a file into SEREBO database.

    A dictionary of items generated will be returned with
    the following keys: (1) DateTimeStamp is the UTC date
    time stamp of this event, (2) Data is the given data
    string to be inserted, (3) UserDescription is the user
    given explanation string for this event, (4) DataHash
    is the hash string of Data, (5) ParentBlockID is the ID
    of the parent block in blockchain, (6)
    ParentDateTimeStamp is the UTC date time stamp of the
    parent block in blockchain (which is also the parent
    insertion event), (7) ParentRandomString is the random
    string generated in parent block in blockchain, (8)
    ParentHash is the hash of parent block in blockchain,
    (9) BlockRandomString is the random string generated
    for current insertion event, and (10) BlockHash is the
    block hash of current insertion event in blockchain.

    @param sdb_object Object: SEREBO database object.
    @param fileapth String: Path of file to log in SEREBO
    black box.
    @param description String: Explanation string for this
    entry event. Default = NA.
    @return: Dictionary of data generated from this event.
    '''
    absPath = absolutePath(filepath)
    if description == 'NA':
        description = ['UserGivenPath:>%s' % str(filepath),
                       'AbsolutePath:>%s' % str(absPath)]
    else:
        description = ['UserGivenPath :> %s' % str(filepath),
                       'AbsolutePath :> %s' % str(absPath),
                       'UserDescription :> %s' % \
                        str(description)]
```

```
        description = ' >> '.join(description)
        fHash = fileHash(absPath)
        rdata = sdb_object.insertData(fHash, description,
            'file')
        return rdata

def searchDatalog(sdb_object, term, field, mode='like'):
    '''!
    Function to search datalog table.

    @param sdb_object Object: SEREBO database object.
    @param term String: Case sensitive search term.
    @param field String: Field name to search.
    @param mode String: Mode of search. Allowable modes are
    'like' and 'exact'. If mode is 'like', wildcards such
    as '_' (matches any single character) and '%' (matches
    any number of characters). Default = 'like'.
    @return: List of datalog rows: [ID, dtstamp, hash,
    data, description]
    '''
    term = str(term)
    field = str(field)
    if mode.lower() == 'exact':
        sqlstmt = """select ID, dtstamp, hash, data,
            description from datalog where %s='%s'""" % \
            field, term)
    if mode.lower() == 'like':
        sqlstmt = """select ID, dtstamp, hash, data,
            description from datalog where %s like
            '%s'""" %\ (field, term)
    result = [row for row in
    sdb_object.cur.execute(sqlstmt)]
    return result

def dateTime(sdb_object):
    '''!
    Function to get a date time string.

    @param sdb_object Object: SEREBO database object.
    @return: Date time string
```

```
    '''
    return sdb_object.dtStamp()

def randomString(sdb_object, length):
    '''!
    Function to get a random string.

    @param sdb_object Object: SEREBO database object.
    @param length Integer: Length of random string to
    generate.
    @return: Random string
    '''
    length = int(length)
    return sdb_object.randomString(length)

def stringHash(sdb_object, dstring):
    '''!
    Function to generate hash for a data string.

    @param dstring String: Data string for hash generation.
    @param sdb_object Object: SEREBO database object.
    @return: Hash
    '''
    return sdb_object.hash(str(dstring))

def gmtime(seconds_since_epoch):
    '''!
    Function to generate a UTC date time stamp string in
    the format of <year>:<month>:<day>:<hour>: <minute>:
    <second>: <microsecond> from seconds since epoch.
    However, microseconds cannot be converted; hence, it is
    given as 00000.

    @param seconds_since_epoch Float: Seconds since epoch.
    @return: UTC date time stamp string
    '''
    seconds_since_epoch = float(seconds_since_epoch)
    now = time.gmtime(seconds_since_epoch)
    now = [str(now.tm_year), str(now.tm_mon),
           str(now.tm_mday), str(now.tm_hour),
```

```
                str(now.tm_min), str(now.tm_sec),
                '00000']
        return ':'.join(now)

def backup(bbpath, backuppath):
    '''!
    Function to backup SEREBO Black Box.

    @param backuppath String: Path for backed-up SEREBO
    black box.
    @param bbpath String: Path to SEREBO black box.
    @return: (absolute bbpath, absolute backuppath)
    '''
    import shutil
    bbpath = absolutePath(bbpath)
    backuppath = absolutePath(backuppath)
    db = connectDB(bbpath)
    db.cur.execute('begin immediate')
    shutil.copyfile(bbpath, backuppath)
    db.conn.rollback()
    return (str(bbpath), str(backuppath))

def dumpTable(sdb_object, tableName, fieldNames, outputfile):
    '''!
    Function to dump table from SEREBO Black Box to CSV
    file.

    @param sdb_object Object: SEREBO database object.
    @param tableName String: Name of table.
    @param fieldNames List: List of fields to dump.
    @param outputfile String: Path of file to write data
    dump.
    @return: (absolute output file path, number of records
    dumped)
    '''
    tableName = str(tableName)
    fieldNames = [str(x) for x in fieldNames]
    fieldNames = ','.join(fieldNames)
    sqlstmt = 'select %s from %s' % (fieldNames, tableName)
```

```
        outputfile = absolutePath(outputfile)
        ofile = open(outputfile, 'w')
        count = 0
        for row in sdb_object.cur.execute(sqlstmt):
        row = [str(d) for d in row]
            row = ','.join(row)
            ofile.write(row + '\n')
            count = count + 1
        ofile.close()
        return (outputfile, str(count))
```

File Name: SEREBODB.PY

```
    '''!
    Secured Recorder Box (SEREBO) Black Box Interface

    Date created: 17th May 2018
    License: GNU General Public License version 3 for academic or not-for-profit use only

    SEREBO is free software: you can redistribute it and/or modify it under the terms of the GNU General Public License as published by the Free Software Foundation, either version 3 of the License, or (at your option) any later version.
    This program is distributed in the hope that it will be useful, but WITHOUT ANY WARRANTY; without even the implied warranty of MERCHANTABILITY or FITNESS FOR A PARTICULAR PURPOSE. See the GNU General Public License for more details.
    You should have received a copy of the GNU General Public License along with this program. If not, see <http://www.gnu.org/licenses/>.
    '''
from datetime import datetime
import hashlib
import random
import os
```

```python
import secrets
import sqlite3
import string
import time

class SereboDB(object):
    '''
    Class representing SEREBO database - the recorder black
    box.
    '''
    def __init__(self, dbpath):
        '''
        Initiation method - connects to SEREBO database. If
        SEREBO database does not exist, this function will
        create the database with the necessary data tables.

        @param bbpath String: Path to SEREBO black box.
        '''
        self.path = dbpath
        self.conn = sqlite3.connect(self.path)
        self.cur = self.conn.cursor()
        self._createTables()

def dtStamp(self):
    '''
    Method to generate a UTC date time stamp string in
    the format of <year>:<month>:<day>:<hour>:<minute>:
    <second>:<microsecond>
    @return: UTC date time stamp string
    '''
    now = datetime.utcnow()
    now = [str(now.year), str(now.month),
           str(now.day), str(now.hour),
           str(now.minute), str(now.second),
           str(now.microsecond)]
    now = ':'.join(now)
    return now

def randomString(self, length=64):
    '''
```

```
    Method to generate a random string, which can
    contain 80 possible characters - abcdefghijklm
    nopqrstuvwxyzABCDEFGHIJKLMNOPQRSTUVWXYZ0123456
    789~!@#$%^&*()<>=+[]?. Hence, the possible number
    of strings is 80**length.

    @param length Integer: Length of random string to
    generate. Default = 64.
    @return: Random string
    '''
    choices = string.ascii_letters + \
              string.digits + \
              '~!@#$%^&*()<>=+[]?'
    x = random.choices(choices, k=int(length))
    return ''.join(x)

def hash(self, data):
    '''!
    Method to generate a series of 12 hashes for a given
    data string, in the format of <MD5>:<SHA1>:<SHA224>:
    <SHA3_244>:<SHA256>:<SHA3_256>:<SHA384>:<SHA3_384>:
    <SHA512>:<SHA3_215>:<Blake 2b>:<Blake 2s>.

    @param data String: Data string to generate hash.
    @return: Hash
    '''
    data = str(data)
    data = bytes(data, 'utf-8')
    x = [hashlib.md5(data).hexdigest(),
         hashlib.sha1(data).hexdigest(),
         hashlib.sha224(data).hexdigest(),
         hashlib.sha3_224(data).hexdigest(),
         hashlib.sha256(data).hexdigest(),
         hashlib.sha3_256(data).hexdigest(),
         hashlib.sha384(data).hexdigest(),
         hashlib.sha3_384(data).hexdigest(),
         hashlib.sha512(data).hexdigest(),
         hashlib.sha3_512(data).hexdigest(),
         hashlib.blake2b(data).hexdigest(),
         hashlib.blake2s(data).hexdigest()]
```

```
        return ':'.join(x)

def _createTables(self):
    '''!
    Private method - used by initialization method to
    generate data tables.
    '''
    now = self.dtStamp()
    # Metadata table
    sql_metadata_create = '''
    create table if not exists metadata (
        key text primary key,
        value text not null);'''
    sql_metadata_insert1 = '''
    insert into metadata (key, value) values
        ('creation_datetimestamp', '%s');''' % (now)
    sql_metadata_insert2 = '''
    insert into metadata (key, value) values
        ('creation_secondstamp', '%s');''' % \
        str(time.time())
    sql_metadata_insert3 = '''
    insert into metadata (key, value) values
        ('blackboxID', '%s');''' % \
        (self.randomString(512))
    sql_notary_create = '''
    create table if not exists notary (
        ID integer primary key autoincrement,
        dtstamp text not null,
        alias text not null,
        owner text not null,
        email text not null,
        notaryDTS text not null,
        notaryAuthorization text not null,
        notaryURL text not null);'''
    # System data table
    sql_systemdata_create = '''
    create table if not exists systemdata (
        ID integer primary key autoincrement,
        dtstamp text not null,
        key text not null,
```

```
      value text not null);'''
# Data log table
sql_datalog_create = '''
create table if not exists datalog (
   ID integer primary key autoincrement,
   dtstamp text not null,
   hash text not null,
   data blob,
   description blob not null);'''
sql_datalog_unique = '''
   create unique index if not exists datalog_unique
   on datalog (dtstamp, hash);'''
# Blockchain table
sql_blockchain_create = '''
create table if not exists blockchain (
   c_ID integer primary key autoincrement,
   c_dtstamp text not null,
   c_randomstring text not null,
   c_hash text not null,
   p_ID integer not null,
   p_dtstamp text not null,
   p_randomstring text not null,
   p_hash text not null,
   data text not null);'''
# Event log table
sql_eventlog_create1 = '''
create table if not exists eventlog (
   ID integer primary key autoincrement,
   dtstamp text not null,
   fID text not null,
   description text not null);'''
sql_eventlog_create2 = '''
create table if not exists eventlog_datamap (
   dtstamp text not null,
   fID text not null,
   key text not null,
   value text not null);'''
# SQL execution
sqlstmt = [sql_metadata_create,
           sql_metadata_insert1,
```

```
                sql_metadata_insert2,
                sql_metadata_insert3,
                sql_notary_create,
                sql_systemdata_create,
                sql_datalog_create,
                sql_datalog_unique,
                sql_blockchain_create,
                sql_eventlog_create1,
                sql_eventlog_create2]
    for statement in sqlstmt:
        try:
            self.cur.execute(statement)
            self.conn.commit()
        except sqlite3.IntegrityError:
            pass

def _insertData1A(self, data, description):
    '''
    Private method - Step 1 of insert data into SEREBO
    black box. Called by insertData method. Step 1 (1)
    gets a UTC date time stamp; (2) formats the
    description by suffixing the description with a 10-
    character random string (80**10 = 1e19 possibility);
    and (3) generates a hash using the UTC date time
    stamp, data and formatted description.

    The difference between _insertData1A() and
    _insertData1B() methods is that _insertData1A()
    method will suffix the description with a 10-
    character random string before using it to generate
    the hash whereas _insertData1B() method uses the
    original (un-suffixed) description in hash
    generation.
    '''
    dtstamp = self.dtStamp()
    DL_data = str(data)
    if description == 'NA' or description == None:
        description = 'NA:' + self.randomString(10)
    else:
        description = str(description) + ':' + \
```

```
                    self.randomString(10)
    DL_hash = self.hash(bytes(dtstamp, 'utf-8') + \
                        bytes(DL_data, 'utf-8') + \
                        bytes(description, 'utf-8'))
    return (dtstamp, DL_data, description, DL_hash)

def _insertData1B(self, data, description):
    '''!
    Private method - Step 1 of insert data into SEREBO
    black box. Called by insertData method. Step 1 (1)
    gets a UTC date time stamp; and (2) generates a hash
    using the UTC date time stamp, data and description
    containing the absolute and relative path to the
    file.

    The difference between _insertData1A() and
    _insertData1B() methods is that _insertData1A()
    method will suffix the description with a 10-
    character random string before using it to generate
    the hash whereas _insertData1B() method uses the
    original (un-suffixed) description in hash
    generation.
    '''
    dtstamp = self.dtStamp()
    DL_data = str(data)
    description = str(description)
    DL_hash = self.hash(bytes(dtstamp, 'utf-8') + \
                        bytes(DL_data, 'utf-8') + \
                        bytes(description, 'utf-8'))
    return (dtstamp, DL_data, description, DL_hash)

def _insertData2(self, dtstamp, DL_data, description,
                 DL_hash, debug):
    '''!
    Private method - Step 2 of insert data into SEREBO
    black box.
    Called by insertData method. Step 2 inserts the
    results from Step 1 into datalog table.
    '''
    sqlstmt = '''insert into datalog (dtstamp, hash,
```

```
                data, description) values (?,?,?,?)'''
        sqldata = (str(dtstamp), str(DL_hash), str(DL_data),
                   str(description))
        self.cur.execute(sqlstmt, sqldata)
        if debug:
            print('Step 1&2: Inserted Data into Data Log
                ...')
            print('Date Time Stamp: %s' % dtstamp)
            print('Inserted Data: %s' % data)
            print('Generated Hash: %s' % DL_hash)

def _insertData3(self, debug):
    '''
    Private method - Step 3 of insert data into SEREBO
    black box. Called by insertData method. Step 3 gets
    data (ID, dtstamp, randomstring, and hash) the
    latest pre-existing block in blockchain table, to be
    used as parent in the next block.
    '''
    sqlstmt = '''select max(c_ID) from blockchain'''
    max_cID = [row
               for row in self.cur.execute(sqlstmt)][0][0]
    if max_cID == None:
        p_ID = 0
        p_dtstamp = '0'
        p_randomstring = \
            'GenesisBlock:SEREBO_MauriceHTLing'
        p_hash = 'TheWord:OmAhHum'
    else:
        sqlstmt = '''select c_ID, c_dtstamp,
            c_randomstring,
            c_hash from blockchain where c_ID = %s''' %\
            str(max_cID)
        data3 = [row for row in
                 self.cur.execute(sqlstmt)]
        p_ID = data3[0][0]
        p_dtstamp = data3[0][1]
        p_randomstring = data3[0][2]
        p_hash = data3[0][3]
    if debug:
```

```
        print('Step 3: Getting Latest Block from
            Blockchain ...')
        print('Parent ID: %s' % p_ID)
        print('Parent Date Time Stamp: %s' % p_dtstamp)
        print('Parent Random String: %s' % \
            p_randomstring)
        print('Parent Hash: %s' % p_hash)
    return (p_ID, p_dtstamp, p_randomstring, p_hash)

def _insertData4(self, p_dtstamp, p_randomstring,
                p_hash, DL_hash):
    '''
    Private method - Step 4 of insert data into SEREBO
    black box. Called by insertData method. Step 4 (1)
    generates a hash from the parent date time stamp,
    parent random string, parent hash, and current data
    hash (from Step 1) as current block hash; (2) and
    generates a 32-character random string (80**32 =
    8e60 possibilities) for the current block.
    '''
    BC_rstr = self.randomString(32)
    hashdata = ''.join([str(p_dtstamp),
                str(p_randomstring),
                str(p_hash), str(DL_hash)])
    BC_hash = self.hash(bytes(hashdata, 'utf-8'))
    return (BC_rstr, BC_hash)

    def _insertData5(self, dtstamp, BC_rstr, BC_hash, p_ID,
                p_dtstamp, p_randomstring, p_hash,
                DL_hash, debug):
    '''
    Private method - Step 5 of insert data into SEREBO
    black box. Called by insertData method. Step 5
    inserts data of the current block into blockchain
    table.
    '''
    sqldata = (str(dtstamp), str(BC_rstr), str(BC_hash),
            str(p_ID), str(p_dtstamp),
            str(p_randomstring), str(p_hash),
            str(DL_hash))
```

```python
        sqlstmt = '''insert into blockchain (c_dtstamp,
            c_randomstring, c_hash, p_ID, p_dtstamp,
            p_randomstring, p_hash, data) values
            (?,?,?,?,?,?,?,?)'''
        self.cur.execute(sqlstmt, sqldata)
        if debug:
            print('Step 5: Insert Data into Blockchain (New
                    Block) ...')
            print('Random String: %s' % BC_rstr)
            print('New Block Hash: %s' % BC_hash)
            print('')

    def _insertData6(self, dtstamp, description,
                    DL_hash, p_hash, BC_hash):
        '''!
        Private method - Step 6 of insert data into SEREBO
        black box. Called by insertData method. Step 6
        records the current data insertion event into
        eventlog tables by recording the date time stamp,
        parent block hash, data hash, and current block
        hash.
        '''
        fID = self.randomString(10)
        sqlstmt = '''insert into eventlog (dtstamp, fID,
            description) values (?,?,?)'''
        sqldata = (str(dtstamp), str(fID), str(description))
        self.cur.execute(sqlstmt, sqldata)
        sqlstmt = '''insert into eventlog_datamap (dtstamp,
            fID, key, value) values (?,?,?,?)'''
        sqldata = [(str(dtstamp), str(fID), 'DataHash',
                    str(DL_hash)),
                   (str(dtstamp), str(fID), 'ParentHash',
                    str(p_hash)),
                   (str(dtstamp), str(fID), 'BlockHash',
                    str(BC_hash))]
        self.cur.executemany(sqlstmt, sqldata)

    def insertData(self, data, description='NA',
                   mode='text', debug=False):
        '''!
```

Method to insert data into SEREBO database. Data will be recorded in datalog table together with the hash of the data. The hash of the data will be logged into blockchain table. This data insertion event will be logged into eventlog tables.

A dictionary of items generated will be returned with the following keys: (1) DateTimeStamp is the UTC date time stamp of this event, (2) Data is the given data string to be inserted, (3) UserDescription is the user given explanation string for this event suffixed with a 64-character random string, (4) DataHash is the hash string of Data, (5) ParentBlockID is the ID of the parent block in blockchain, (6) ParentDateTimeStamp is the UTC date time stamp of the parent block in blockchain (which is also the parent insertion event), (7) ParentRandomString is the random string generated in parent block in blockchain, (8) ParentHash is the hash of parent block in blockchain, (9) BlockRandomString is the random string generated for current insertion event, and (10) BlockHash is the block hash of current insertion event in blockchain.

@param data String: Data to be inserted.
@param description String: Explanation string for this entry event. Default = NA.
@param mode String: Type of data to insert. Allowable modes are 'text' (description text is suffixed with a 10-character random string), 'ftext' (description text is not suffixed with a 10-character random string) and 'file' (for file hash logging). Default = 'text'.
@param debug Boolean: Flag to print out debugging statements.
@return: Dictionary of data generated from this event.
'''
# Step 1: Preparing data
if mode.lower() == 'text':

```
            (dtstamp, DL_data, description, DL_hash) = \
                self._insertData1A(data, description)
        elif mode.lower() == 'file':
            (dtstamp, DL_data, description, DL_hash) = \
                self._insertData1B(data, description)
        elif mode.lower() == 'ftext':
            (dtstamp, DL_data, description, DL_hash) = \
                self._insertData1B(data, description)
        # Step 2: Insert data into datalog
        self._insertData2(dtstamp, DL_data, description,
                          DL_hash, debug)
        # Step 3: Get latest block in blockchain
        (p_ID, p_dtstamp, p_randomstring, p_hash) = \
            self._insertData3(debug)
        # Step 4: Prepare data for blockchain insertion
        (BC_rstr, BC_hash) = self._insertData4(p_dtstamp,
                                               p_randomstring,
                                               p_hash,
                                               DL_hash)
        # Step 5: Insert data into blockchain
        self._insertData5(dtstamp, BC_rstr, BC_hash, p_ID,
                          p_dtstamp, p_randomstring, p_hash,
                          DL_hash, debug)
        # Step 6: Insert event into eventlog
        self._insertData6(dtstamp, description,
                          DL_hash, p_hash, BC_hash)
        # Step 7: Commit
        self.conn.commit()
        # Step 8: Return data
        return {'DateTimeStamp': dtstamp,
                'Data': data,
                'UserDescription': description,
                'DataHash': DL_hash,
                'ParentBlockID': p_ID,
                'ParentDateTimeStamp': p_dtstamp,
                'ParentRandomString': p_randomstring,
                'ParentHash': p_hash,
                'BlockRandomString': BC_rstr,
                'BlockHash': BC_hash}
```

Code Files for SEREBO Notary

File Name: SERVICES.PY

```
'''!
Secured Recorder Box (SEREBO) Notary Services

Date created: 19th May 2018

License: GNU General Public License version 3 for
academic or not-for-profit use only

SEREBO is free software: you can redistribute it and/or
modify it under the terms of the GNU General Public
License as published by the Free Software Foundation,
either version 3 of the License, or (at your option) any
later version.
This program is distributed in the hope that it will be
useful, but WITHOUT ANY WARRANTY; without even the implied
warranty of MERCHANTABILITY or FITNESS FOR A PARTICULAR
PURPOSE.  See the GNU General Public License for more
details.
You should have received a copy of the GNU General
Public License along with this program. If not, see
<http://www.gnu.org/licenses/>.
'''

from datetime import datetime
import hashlib as h
import random
import string

from gluon.tools import import Service

service = Service()

def call():
    '''!
    Function to enable web services in Web2Py.
```

```python
    '''
    session.forget()
    return service()

@service.xmlrpc
def now():
    '''!
    Function to generate a UTC date time stamp string in
    the Format of <year>:<month>:<day>:<hour>: <minute>:
    <second>:<microsecond>

    @return: UTC date time stamp string
    '''
    dt = datetime.utcnow()
    x = [str(dt.year), str(dt.month),
         str(dt.day), str(dt.hour),
         str(dt.minute), str(dt.second),
         str(dt.microsecond)]
    return ':'.join(x)

@service.xmlrpc
def randomString(length=16):
    '''!
    Function to generate a random string, which can contain
    80 possible characters - abcdefghijklmnopqrstuvwxyz
    ABCDEFGHIJKLMNOPQRSTUVWXYZ0123456789~!@#$%^&*()<>=+[]?.
    Hence, the possible number of strings is 80**length.

    @param length Integer: Length of random string to
    generate. Default = 16.
    @return: Random string
    '''
    choices = string.ascii_letters + \
              string.digits + \
              '~!@#$%^&*()<>=+[]?'
    x = [random.choice(choices)
         for i in range(int(length))]
    return ''.join(x)

@service.xmlrpc
```

```
def register_blackbox(blackboxID, owner, email,
          architecture, machine, node,
          platform, processor):
    '''
    Function to register SEREBO Black Box with SEREBO
    Notary.

    @param blackboxID String: ID of SEREBO black box -
    found in metadata table in SEREBO black box database.
    @param owner String: Owner's or administrator's name.
    @param email String: Owner's or administrator's email.
    @param architecture String: Architecture of machine -
    from platform library in Python Standard Library.
    @param machine String: Machine description - from
    platform library in Python Standard Library.
    @param node String: Machine's node description - from
    platform library in Python Standard Library.
    @param platform String: Platform description - from
    platform library in Python Standard Library.
    @param processor String: Machine's processor
    description - from platform library in Python Standard
    Library.
    @returns: (Notary authorization code, Date time stamp
    from SEREBO Notary)
    '''
    dtstamp = now()
    notaryAuthorization = str(randomString(256))
    notabase.registered_blackbox.insert(datetimestamp=dtstamp,
                        blackboxID=str(blackboxID),
                        owner=str(owner),
                        email=str(email),
                        architecture=str(architecture),
                        machine=str(machine),
                        node=str(node),
                        platform=str(platform),
                        processor=str(processor),
                        notaryAuthorization=notaryAuthorization)
    eventText = ['SEREBO Black Box Registration',
                 'Date Time Stamp: %s' % dtstamp,
                 'Black Box ID: %s' % blackboxID,
```

```
                   'Owner: %s' % owner,
                   'Email: %s' % email,
                   'Notary Authorization: %s' % \
                      notaryAuthorization]
    eventText = ' | '.join(eventText)
    notabase.eventlog.insert(datetimestamp=dtstamp,
                             event=eventText)
    return (notaryAuthorization, dtstamp)

@service.xmlrpc
def hash(dstring):
    '''!
    Function to generate a series of 6 hashes for a given
    data string, in the format of <MD5>:<SHA1>:<SHA224>:
    <SHA256>: <SHA384>:<SHA512>.

    @param dstring String: String to generate hash.
    @return: Hash
    '''
    dstring = str(dstring)
    x = [h.md5(dstring).hexdigest(),
         h.sha1(dstring).hexdigest(),
         h.sha224(dstring).hexdigest(),
         h.sha256(dstring).hexdigest(),
         h.sha384(dstring).hexdigest(),
         h.sha512(dstring).hexdigest()]
    return ':'.join(x)

@service.xmlrpc
def checkBlackBoxRegistration(blackboxID,
                              notaryAuthorization):
    '''!
    Function to check for SEREBO Black Box registration.

    @param blackboxID String: ID of SEREBO black box -
    found in metadata table in SEREBO black box database.
    @param notaryAuthorization String: Notary authorization
    code of SEREBO black box (generated during black box
    registration - found in metadata table in SEREBO black
    box database.
```

```
    @return: True if SEREBO Black Box is registered; False
    if SEREBO Black Box is not registered
    '''
    if notabase(notabase.registered_blackbox.blackboxID ==
        \ blackboxID) \
    (notabase.registered_blackbox.notaryAuthorization == \
        notaryAuthorization).count():
        return True
    else:
        return False

@service.xmlrpc
    def notarizeSereboBB(blackboxID, notaryAuthorization,
                        dtstampBB, codeBB):
    ''''!
    Function to notarize SEREBO Black Box with SEREBO
    Notary.
    @param blackboxID String: ID of SEREBO black box -
    found in metadata table in SEREBO black box database.
    @param notaryAuthorization String: Notary authorization
    code of SEREBO black box (generated during black box
    registration - found in metadata table in SEREBO black
    box database.
    @param dtstampBB String: Date time stamp from SEREBO
    black box.
    @param codeBB String: Notarization code from SEREBO
    black box.
    @returns: (Date time stamp from SEREBO Notary,
    Notarization code from SEREBO Notary, Cross-Signing
    code from SEREBO Notary)
    '''
    blackboxID = str(blackboxID)
    notaryAuthorization = str(notaryAuthorization)
    dtstampBB = str(dtstampBB)
    codeBB = str(codeBB)
    dtstampNS = now()
    codeNS = str(randomString(32))
    codeCommon = hash(codeBB + codeNS)
    notabase.notarize_blackbox.insert(blackboxID=
    blackboxID,
```

```
            notaryAuthorization=notaryAuthorization,
                              dtstampBB=dtstampBB,
                              dtstampNS=dtstampNS,
                              codeBB=codeBB,
                              codeNS=codeNS,
                              codeCommon=codeCommon)
    eventText = ['SEREBO Black Box Notarization',
                 'Success',
                 'Date Time Stamp: %s' % dtstampNS,
                 'Black Box ID: %s' % blackboxID,
                 'Notary Authorization: %s' % \
                    notaryAuthorization,
                 'Black Box Code: %s' % codeBB,
                 'Notary Code: %s' % codeNS,
                 'Cross-Signing Code: %s' % codeCommon]
    eventText = ' | '.join(eventText)
    notabase.eventlog.insert(datetimestamp=dtstampNS,
                             event=eventText)
    return (dtstampNS, codeNS, codeCommon)

@service.xmlrpc
def checkNotarizeSereboBB(blackboxID, notaryAuthorization,
                          BBCode, NCode, CommonCode):
    '''!
    Function to check for SEREBO Black Box notarization by
    SEREBO Notary.

    @param blackboxID String: ID of SEREBO black box -
    found in metadata table in SEREBO black box database.
    @param notaryAuthorization String: Notary authorization
    code of SEREBO black box (generated during black box
    registration - found in metadata table in SEREBO black
    box database.
    @param BBCode String: Notarization code from SEREBO
    Black Box.
    @param NCode String: Notarization code from SEREBO
    Notary.
    @param CommonCode String: Cross-Signing code from
    SEREBO Notary.
    @return: True if notarization is found; False if
```

```
    notarization is not found.
    '''
    if notabase(notabase.notarize_blackbox.blackboxID == \
        blackboxID) \
        (notabase.notarize_blackbox.notaryAuthorization == \
        notaryAuthorization) \
        (notabase.notarize_blackbox.codeBB == BBCode) \
        (notabase.notarize_blackbox.codeNS == NCode) \
        (notabase.notarize_blackbox.codeCommon == \
        CommonCode).count():
        return True
    else:
        return False
```

File Name: SEREBO_NOTABASE.PY

```
    '''!
    Secured Recorder Box (SEREBO) Notary Database

    Date created: 19th May 2018

    License: GNU General Public License version 3 for
academic or not-for-profit use only

    SEREBO is free software: you can redistribute it and/or
modify it under the terms of the GNU General Public
License as published by the Free Software Foundation,
either version 3 of the License, or (at your option) any
later version.
    This program is distributed in the hope that it will be
useful, but WITHOUT ANY WARRANTY; without even the implied
warranty of MERCHANTABILITY or FITNESS FOR A PARTICULAR
PURPOSE. See the GNU General Public License for more
details.
    You should have received a copy of the GNU General
Public License along with this program. If not, see
<http://www.gnu.org/licenses/>.
    '''
```

```python
notabase = SQLDB('sqlite://serebo_notabase.sqlite')
'''
Table registered_blackbox is to store registration data
of SEREBO Black Box.
'''
notabase.define_table('registered_blackbox',
    SQLField('datetimestamp', 'text'),
    SQLField('blackboxID', 'text', unique=True),
    SQLField('owner', 'text'),
    SQLField('email', 'text'),
    SQLField('architecture', 'text'),
    SQLField('machine', 'text'),
    SQLField('node', 'text'),
    SQLField('platform', 'text'),
    SQLField('processor', 'text'),
    SQLField('notaryAuthorization', 'text'))

'''
Table notarize_blackbox is to store notarization data
when SEREBO Black Box requests for SEREBO Notary's
notarization.
'''
notabase.define_table('notarize_blackbox',
    SQLField('blackboxID', 'text'),
    SQLField('notaryAuthorization', 'text'),
    SQLField('dtstampBB', 'text'),
    SQLField('dtstampNS', 'text'),
    SQLField('codeBB', 'text'),
    SQLField('codeNS', 'text'),
    SQLField('codeCommon', 'text'))

'''
Table eventlog is to keep a record of notable events
in SEREBO Notary.
'''
notabase.define_table('eventlog',
    SQLField('datetimestamp', 'text'),
    SQLField('event', 'text'))
```

REFERENCES

Bik, E. M., Fang, F. C., Kullas, A. L., Davis, R. J. & Casadevall, A. (2018). Analysis and Correction of Inappropriate Image Duplication: the Molecular and Cellular Biology Experience. *Molecular Biology of the Cell*, *38*, e00309-18.

Boneh, D. & Boyen, X. (2006). *On the impossibility of efficiently combining collision resistant hash functions*. (Springer), pp. 570–583.

Broder, A. & Mitzenmacher, M. (2001). *Using multiple hash functions to improve IP lookups*. (IEEE), pp. 1454–1463.

Dolinski, K. & Troyanskaya, O. G. (2015). Implications of Big Data for cell biology. *Molecular Biology of the Cell*, *26*, 2575–2578.

Ling, M. H. (2013). NotaLogger: Notarization Code Generator and Logging Service. *The Python Papers*, *9*, 2.

Ling, M. H. (2018a). SEcured REcorder BOx (SEREBO) Based on Blockchain Technology for Immutable Data Management and Notarization. *MOJ Proteomics & Bioinformatics*, *7*, 169–174.

Ling, M. H. (2018b). A Cryptography Method Inspired by Jigsaw Puzzles. In *Current STEM, Volume 1*, (Nova Science Publishers, Inc.), pp. 129–142.

Pierce, A. (2010). The Evolution of the Airplane Black Box. *Tech Directions*, *70*, 12.

Pierro, M. D. (2008). *Web2Py: Enterprise Web Framework* (New Jersey, USA: John Wiley & Sons, Inc.).

Rasjid, Z. E., Soewito, B., Witjaksono, G. & Abdurachman, E. (2017). A review of collisions in cryptographic hash function used in digital forensic tools. *Procedia Computer Science*, *116*, 381–392.

Suwinski, P., Ong, C., Ling, M. H. T., Poh, Y. M., Khan, A. M. & Ong, H. S. (2019). Advancing Personalized Medicine Through the Application of Whole Exome Sequencing and Big Data Analytics. *Frontiers in Genetics*, *10*, 49.

Zheng, Z., Xie, S., Dai, H., Chen, X. & Wang, H. (2017). An Overview of Blockchain Technology: Architecture, Consensus, and Future Trends. In *2017 IEEE International Congress on Big Data (BigData Congress)*, pp. 557–564.

Date Submitted: February 20, 2019. Date Accepted: March 29, 2019.

Chapter 4

THE CHANGES WHILE STUDYING IN THE DIFFERENT COUNTRIES AND REFLECTIONS TO THE MOMENTS

Gyun Tae Bae*
Department of Applied Sciences, Northumbria University, Newcastle, UK
School of Life Sciences, Management Development Institute of Singapore, Singapore

ABSTRACT

This autobiography begins with my background started in South Korea. It also includes my chronological experiences and reflection from living and studying in South Korea, Philippines and Singapore. As I had studied in the different countries, there were various moments that made me improve, rethink and develop myself. The autography describes my changes over the years to adapt to the new environments.

---

* Corresponding author E-mail: rbsghkdwpp@gmail.com.

BACKGROUND AND STUDIES IN SOUTH KOREA

I was born on 2nd December 1997 in South Korea. My mother's name is Mun Hee Jung and father is Bae Jin Ha. During that time, my mother's job was a high school teacher as well as tutor. In addition to my parents' jobs, my father was a loan shark. What is a loan shark exactly? Loan shark is a person who lends money to others and get it back with high interest (Mayer, 2013). In other words, we had a lot of money at that time. As time went by, my mother realized that earning money in that way would lead to a bad outcome or a curse to me in the future. Because of my mother's concerns for me, my father decided to retire from his job to take care of me. Everything was fine as my mother earned money more than my father.

In 2000, our family was in financial problem due to bankruptcy. The bankruptcy occurred because my father made a bad decision on an investment stock. My father used the money, which was from my mother, to invest in stock. However, the price of stock was continuously declining, and my father ended up failing. This stock failure led my father to borrow the money from a loan shark to invest in other companies' stocks. Unfortunately, my father became a credit delinquent due to another failure of an investment stock. Because of that, I was taken care of by my paternal and maternal grandmothers until I was about 8 years old.

Even though we were bankrupt, my parents decided to push me to a kindergarten. Firstly, my parents wondered who could take care of me while they were going to work in order to repay the loan cause by my father. It was my paternal grandmother who was willing to take care of me. So, I lived with my paternal grandmother and my parents gave her money for my kindergarten payment as well as her efforts. The funny thing that you must not forget is that she was my father's birth mother. However, my paternal grandmother used the money, given by my parents, for a rotating-credit association. What my paternal grandmother did was deceive my parents by convincing them that the money was not enough for her to cover the cost of taking care of me and the kindergarten payment. With her continuous lies, my paternal grandmother was caught by my parents. I was then sent to my maternal grandmother when I was 5 years old.

My maternal grandmother is a tightwad compared to my paternal grandmother. I often could not eat snacks, ice-cream and street food. So, I missed my paternal grandmother whenever I failed to get to eat those delicious food by crying. It is very shameful that I had to cry to get attention from others and my maternal grandmother for snacks. But I was a young boy and crying was the most powerful and efficient way to achieve success. Other than that, I am grateful for the time I spent with my maternal grandmother. As I got older, I came to realize why my maternal grandmother is a tightwad. My maternal grandmother was millionaire but due to her father's foolish actions, she became economically underprivileged in a short period. For now, I consider that money is rotational – everyone will have a certain period of time as an opportunity to earn money to improve his or her life, or encounter more hardship and end up being worse than before.

When I turned 8 years old, I was taken care of by my parents. My parents were economically stable at that time. It was also time for me to attend elementary school. I went to Yangji Elementary School. I faced much difficulties with subjects such as Mathematics and English. These two subjects made me think in a creative way that I could use to cheat my parents. All the students who study at Yangji Elementary School had to get his or her parents' signature on the examination result to confirm that it was checked. For my poor Mathematics and English results, I forged my parents' signature and I told them that I got above 90 marks out of 100. In South Korea, elementary school is 6 years of study. After which, I graduated and went on to middle school when I was 14 years old.

This was at that time when all my cheats were revealed to my parents as a result of the middle school system where the results of students' exam paper were released online where the parents can have easily access. Since I was not serious in studying, I did not listen to the middle school executives. In other words, I did not know my parents would be able to access my bad results. Of course, I deceived them using the same methods. Initially, I was happy that I succeeded and got additional allowance from my parents. However, right after that day, my parent knew what I had done. All my subject scores were around 30 to 40 marks out of 100. I was scolded and not allowed to play a single computer game. However, the way that they scolded and punished me did not really make me change. So, my parents approached me and made a deal with me. The deal was that if I was able to get 80 marks out of 100 for every subject, my parents would grant me my wish. My wish was to be a building owner so I studied hard. This was especially for two subjects, Mathematics and English, and I memorized everything to achieve my goal. During the Mathematics examination, I could successfully apply the memorized formula to the questions. Unluckily, I had a problem in English, which was only the subject that I could not get more than 80 marks. Usually, the questions in English examination will be asked in Korean. However, every question was in English this time round and I was totally messed up. Obviously, my wish was not granted by my parents. For that, I came to realize that having a goal can change a person.

LIFE WITH STUDIES IN THE PHILIPPINES

My goal was to be able to speak English fluently. In order to achieve my goal, I decided to study in Tagaytay City of the Philippines when I was 15 years old. So, I took a semester off to work on this goal. I realized that I did not even know the basics of English such as the alphabets. I started to learn everything from the beginning. It was difficult for me to understand and communicate with everyone in the Philippines until about 6 months later. It was really surprised that I could gradually speak in English with others, little by little. Additionally, my English examination results were better than before.

All the people I met in Tagaytay City of the Philippines were 10 to 15 years older than me. I always had to be polite to them as for South Korean, perceiving as being polite to others is very important. So, I was always yielding for basically everything. In other words, most of the people there though of me as a kind person whom is always alright to be influenced. It was fine since I learnt how to deal with other people and always being polite, kind and nice to them can be to my disadvantages. So, I decide that

I should not be with those people and focus on myself. As a result of that change, I am stronger than before, and it was much easier to meet and distinguish among many other people.

My plan was to study for another 6 months before going back home. All of the sudden, my guardian suggested to me to study at a high school in Philippines, which is Mater Dei Academy High School, in 2013. This school is a Catholic school where most of the teachers are sisters (Mater Dei Academy, 2013). The subjects taught were the same as usual schools. However, there was one subject, called Bible Studies, that I struggled in. Bible Studies was about Gods' sayings and its translation to reality. One of the most unforgettable lectures was about freedom. The definition of freedom is one's right to act and speak (Domingo, 2014). Additionally, everyone wants to have it. Because of what my sister said, the word freedom seemed not always good to hear. The sister said whenever someone gets freedom, the other must be suffering from his or her freedom (Bielefeldt, 2012). Somehow, I had a grateful time and enjoyed studying in Mater Dei Academy High School.

Turning to fourth year or last year of high school in March 2015, I moved to another high school called Tropical Innovative School of Excellence (TISE), located in Tagaytay City of the Philippines. The reason was that I was not much of a religious student to understand the curriculum system at Mater Dei Academy High School. Soon after, I met several Korean friends whom I have kept in touch for more than 4 years. That time, I was really lazy to wake up early in the morning just to attend the classes. I did not care about what I had to do. If I was not mistaken, I went to the school 3 times a week However, it was not that difficult to pass examinations and I achieved my academic goal of being within the top 10 students. This pushed me to want to study further. One of the subjects, that made me think in different ways, was Science. For me, Science is something attractive but difficult to get in deeply. In order to get into Science deeper than before, I decided to pursue a course in Biotechnology and I came to know Management Development Institute of Singapore (MDIS) campus has a course in Biotechnology awarded by Northumbria

University (Northumbria University, N.D.). So, I made plans to go to Singapore, step by step.

ACHIEVEMENT TO THE GOAL STEP BY STEP

1. Understand and admit areas of weakness to improve or develop.
2. Have resolutions targeted towards the weak points.
3. Make a purpose to proceed.
4. Must not move emotionally and neglect the processes of the goal.

LIFE WITH STUDIES IN SINGAPORE

Before I attend the degree course in Biotechnology, I had to complete a Diploma in Biomedical Science. One of the major requirements required by MDIS coordinator was International Early Learning Study (IELS) score, which must be above 6.0, and results of my high school. Luckily, I did not have to take IELS due to the fact that I had taken compulsory subjects in English. I was offered with tuition grant. For that, I had SDG 900 discounted off my diploma tuition fee. I started the program in August 2015. During my Diploma in Biomedical Science, I had a difficult time as I had to wake up early every day to attend classes for 6 hours. Honestly, I still have phobia with number so I hate Mathematics. I also had to take middle term test for every module and that made me plan my schedule properly as I was eager to achieve my goals. I came to face my limitations on a constant basis – almost once every fortnight. However, I strived on in face and managed to overcome each limitation. According to my classmates, I am able to memorize a lot of notes related to the modules within short time – something I did not know. I realized that once I understand the theories, contents and mechanism, I could get into the interesting aspects of Science much easier. I can tell that life must begin with interest in something before a plan can be formulated and followed. Without interest, it would be difficult to make and follow any plans.

In February 2016, I had finished the course of Diploma in Biomedical Science with distinction. To get a distinction certificate, the average of every module must be higher than 70 marks out of 100. It was quite difficult since 12 modules were taught within 6 months. So, I was really satisfied with myself. In April 2016, I had joined the Biotechnology degree course offered by Northumbria University, which was my plan that I made in the Philippines. The duration of Biotechnology course is almost 4 years which is longer than any other courses provided in MDIS. As a graduate from Diploma in Biomedical Science, I could not think of any course which is not related to Science. I was not as busy as during my diploma course. Maybe it was because there were no morning classes. Every module within Biotechnology course was much more in depth compared to modules in diploma.

Another goal I had wanted to accomplish was to get first honours on my degree graduate certificate. I thought it would be very easy to attain it. As I studied more, I had to admit my limitations again. The main limitation was my ability to understanding on the research papers published by individual investigators. I did not know I had to go through research papers related to various topics within the Biotechnology modules. Another problem was trying to critique the research papers and possibly come to different conclusions than that of the researchers. I realized that there are no right answers in terms of Science but pieces of evidence to support the truth emerges as time goes by.

During the entire first year of the Biotechnology course, every module was repetitive for the students to remind them of the basics of certain topics. I learnt that Science requires a lot of examples and cases to understand profoundly. The result of first year was not included in calculating honour classification. However, I worked harder than during my diploma to fulfil something that I did not know. I also met new friends like Aaron (Indonesian) and Lau (Malaysian) whom I really envied. The reason was that they are smart, and they seemed to understand everything within a short period. As I got deeper into my studies, I found what the real challenge was, and I wanted to fulfil. It was that I wanted to win the competition between me and myself – a competition to advance myself. I realized that before I compete with others, I should have won the competition against my previous self. By winning my internal competition, I believe that I had become stronger and that gives me a lot of opportunities to prepare for alternatives including potentially bad situations in the future. In my perceptive, this step is very important in Science.

I studied in the same way for second year of my Biotechnology course. I had a lot of laboratory reports to submit other than taking examinations at that time. Honestly, I prefer to submit laboratory reports rather than to take examinations. For me, if the result from my examination was not what I had expected, I would think I had studied the module in a wrong way. However, I tend to get good grades in the laboratory reports. Luckily, I also got good grades in examinations. I confirmed to myself that the method I used to study was correct and I decided to use the same method in my studies but with more passion. In a way, I was experimenting on myself. With that, I was rather sure to achieve my goal – first class honours on my graduate certificate.

LIFE DURING GRADUATING YEAR AND MY FINAL YEAR PROJECT

Turing into final year or graduating student, I was thinking how to get a job in Singapore other than to change the way I have studied. In 2018, the first term of the last year of the Biotechnology course, the results of my examination were not what I expected. Every score I got was around 50-60 which may make it difficult in getting first honours on my graduate certificate. So, I had to focus on myself again to change the way I used to study. It was very difficult to realize which steps were wrong. I had to admit my limitation that I was not better than others. In other words, I wanted to achieve too much in too short a time. When it was the last term of my course, I decided to challenge my limitation again. It was to relax myself more than before.

Being strict with myself or relaxing more have both good and weak points. If there is something wrong that I had encountered, it is obvious that I will be disappointed with myself but have hard time to look back the steps I have proceeded. Relaxing more may not develop, enhance and improve myself as quickly but I am sure that I will not have hard time to consider which steps that I had proceeded were wrong. Northumbria University includes one of the systems that students must carry out individual investigation on a given topic. I chose Dr Sunesh as my supervisor because my senior told me that I must be independent. So, I tried to relax myself and be independent at the same time.

Final year project, which is very important to support my goal, was investigation of actinomycetes in terms of biodiversity and physical, chemical and antimicrobial characteristics. Actinomycetes are recognized as biotechnologically valuable bacteria with potential for secondary metabolites production, including antibiotics, pesticides, medicines, toxins and animal and plant growth factors, worldwide in the area of research for many years (Singh, 2013).

Actinomycetes are known as producers of odorous compounds such as geosmin and 2-methylisoborneol. These organic compounds lead to pleasant earthy-musty odour soils (moist soils) and problem in drinking water (Asquith et al., 2013; Liu et al., 2015; Srinivasan and Sorial, 2011). These compounds has an unpleasant taste and odour that consumers cannot accept, and the industry pay a high cost for treatment (Asquith et al., 2013; Lawton, 2003).

Several researchers had investigated whether actinomycetes contributed to the water problem by the production of geosmin and 2-methylisoborneol. However, actinomycetes are not the only geosmin and 2-methylisoborneol producing bacteria as *Anabaena* spp. and cyanobacteria are also able to produce these compounds (Asquith et al., 2013; Lee et al., 2017). Moreover, there is positive correlation between *Anabaena* spp. and concentration of geosmin and 2-methylisoborneol (Park et al., 2014). However, because of the contradiction that algae produce low biomass during cold season, actinomycetes may be considered to contribute to unpleasant taste and odour water (Lee et al., 2011).

With these interesting facts about actinomycetes, I started to collect soils from South Korea and Singapore to study the actinomycetes. The experiment was simple but laborious. I isolated 6 actinomycetes from the sample from Singapore and rest of them from South Korea. All of them had different characteristics and can be classified into specific categories based on the obtained results. Final year project was a practice to work independently in the society (Todd et al., 2004).When it was close to deadline, I could feel that my degree course was getting over and I wanted to work in Singapore.

LIFE AFTER THE SEMESTERS OFF AND WHAT I FEEL ON MY PREVIOUS STUDIES

Since 18th February 2019, I have waited for the results of my goal and had submitted my curriculum vitae to the companies where I want to work. While waiting for both of that, it is very hard to do nothing. Recently, I am easily disappointed with myself for not receiving any reply from companies. This situation made me wanted to find out more solutions. I believe there is nothing certain for now as I am not clear of my possible paths into the future. I cannot know which route will make me successful unless I choose. However, this problem makes me consider of another plan which is to pursue a Master in Biomedical Engineering in Germany. These experiences are common to everyone and nobody knows the right answer for it.

I have followed the same steps for both my studies in the Philippines and Singapore. I strongly believe these steps are very important when study abroad. The difference during my studies in the Philippines and Singapore was age. I was still young when I studied in the Philippines. Obviously, the way I thought was not mature. In other words, the goal was immature. Additionally, it was difficult to follow through the steps. At that age, I could not think of appreciating my parents for their supports and offers. As I further my studies abroad especially in Singapore, the way I think is totally different compared to previously. I realized that I cannot achieve what I have without support from my parents. Every parent in the world has lived as a manure to facilitate in germinating their seeds which are offspring. Although I am not forced to follow through the steps, the steps are essential to my life. One more thing I have realized is that everything has its price. Honestly, my life in Singapore is more distressful compare to the Philippine due to the fact that I am more matured and had encountered a lot of obstacles along the way.

As I have studied abroad for more than 10 years, there are many situations I have faced with and I believe that these situations made me think in better ways. I am not too bothered with interpersonal relationships and friendships. People who want to be with me will not leave and people who want to leave will still leave me despite me do my best to be with them. Additionally, I will be able to distinguish people, who are on my side, through painful situations. I strongly believe that everyone will face such situations and his or her consequence is dependent on how he or she uses and overcomes the situations.

REFERENCES

Asquith, E., Evans, C., Geary, P., Dunstan, R. and Cole, B. (2013). The role of Actinobacteria in taste and odour episodes involving geosmin and 2-methylisoborneol in aquatic environments. *Journal of Water Supply: Research and Technology-Aqua*, 62(7), pp. 452-467.

Bielefeldt, H. (2012). Freedom of Religion or Belief--A Human Right under Pressure. *Oxford Journal of Law and Religion*, 1(1), pp. 15-35.

Domingo, R. (2014). A right to religious and moral freedom? *International Journal of Constitutional Law*, 12(1), pp. 226-247.

Lawton, L. (2003). The destruction of 2-methylisoborneol and geosmin using titanium dioxide photocatalysis. *Applied Catalysis B: Environmental*, 44(1), pp. 9-13.

Lee, G., Kim, Y., Kim, M., Oh, S., Choi, I., Choi, J., Park, J., Chong, C., Kim, Y., Lee, K. and Lee, C. (2011). Presence, molecular characteristics and geosmin producing ability of Actinomycetes isolated from South Korean terrestrial and aquatic environments. *Water Science and Technology*, 63(11), pp. 2745-2751.

Lee, J., Rai, P., Jeon, Y., Kim, K. and Kwon, E. (2017). The role of algae and cyanobacteria in the production and release of odorants in water. *Environmental Pollution*, 227, pp. 252-262.

Liu, H., Pan, D., Zhu, M. and Zhang, D. (2015). Occurrence and Emergency Response of 2-Methylisoborneol and Geosmin in a Large Shallow Drinking Water Reservoir. *CLEAN - Soil, Air, Water*, 44(1), pp. 63-71.

Mayer, R. (2013). Loansharking. *The Encyclopedia of Criminology and Criminal Justice*, pp. 1-5.

Mater Dei Academy, (2013). *Introduction and Philosophy*. [Online] Available from: http://www.materdeiacademy.us/wp-content/uploads/2013/08/Family-Handbook-2013-2014-8-23-13.pdf. [Accessed on 10 March 2019].

Northumbria University, (N.D.). *Management Development Institute Singapore (MDIS) Programmes*. [Online] Available from: https://www.northumbria.ac.uk/ international/ international-partners/ health-life- Sciences- international- partners/ management- development-institute-singapore-mdis/. [Accessed on: 13 March 2019].

Park, T., Yu, M., Kim, H., Cho, H., Hwang, M., Yang, II., Lee, J., Lee, J. and Kim, S. (2014). Characteristics of actinomycetes producing geosmin in Paldang Lake, Korea. *Desalination and Water Treatment*, 57(2), pp. 888-899.

Singh, H. (2013). Diversity and Versatility of Actinomycetes and its Role in Antibiotic Production. *Journal of Applied Pharmaceutical Science*, 3 (8 Suppl 1), pp. S83-S94.

Srinivasan, R. and Sorial, G. A. (2011). Treatment of Taste and Odor Causing Compounds 2-Methylisoborneol and Geosmin in Drinking Water: a Critical Review. *J Environ Sci (China)*, 23(1), pp. 1-13.

Todd, M., Bannister, P. and Clegg, Sue. (2004). Independent inquiry and the undergraduate dissertation: perceptions and experiences of final-year social Science students. *Assessment & Evaluation in Higher Education*, 29(3), pp. 336-354.

Date Submitted: March 20, 2019. Date Accepted: April 02, 2019.

In: Current STEM. Volume 2
Editor: Maurice HT Ling

ISBN: 978-1-53616-042-0
© 2019 Nova Science Publishers, Inc.

Chapter 5

MY MEMOIR BASED ON 4 YEARS IN SINGAPORE

Jung Hwan Kim[*]

Department of Applied Sciences, Northumbria University,
Newcastle, UK
School of Life Sciences, Management Development Institute
of Singapore, Singapore

ABSTRACT

In this essay, my background in South Korea, what I have done and what I have felt about university life in Singapore is narrated. I will also describe the valuable experience of studying in Singapore for 4 years, from the diploma to honours. In the last year, with the final year project, it is reasonable to say that it was a turning point of my life. This reflection made me look back into my past and be reminded of the good things that are unforgettable and bad that I regret. Studying in Singapore have brought me a great changes and development, as well as self-reflection. The last 4 years are the best part of my life and I always appreciate everyone whom had supported me for my studies.

---

[*] Corresponding Author's E-mail: joseph266394@gmail.com..

BACKGROUND

My parents are passionate about education and it affects my life substantially. My elder sister was also an overseas student and had finished her degree in United Kingdom. However, I was a conservative person and always dreamt to stay in my country. The educational methods in Korea can best be described as cram education; in other words, students do rote learning even in Mathematics. Due to this, they generally lack applicative and creative aspects. Most students think that it is wrong if their thoughts are different to others. I was also one of them. However, these years overseas had changed me and I am reminded of the proverb: *"A frog in a well knows nothing of the ocean"* (Morck and Nakamura, 2005). When I was in elementary school, I learnt Chemistry I & II and various field of Mathematics. Many high school students learned them to attend a student-academic conference, called Olympiad and I got 3 bronze medals when I was 11 years old. However, early education was a double-edged sword for me. While I had learnt beyond what was expected of me, it affected me negatively. By the time I had completed elementary education and moved into middle school, I had already learnt most of the necessary topics every other student was trying to learn. In tests, I was able to get decent or high marks even though I never study and as a result, I became lazy. The problem was that bad habits are not easily remedied; for example, it is easy to get fat but difficult to lose weight. I tried to change this habit from 16 years old at high school, through staying in high school dormitory with other students.

In contrast with most of other Korean students in Singapore, I graduated from Korean middle school and high school. There are many Korean students who spent their teenager years in English-speaking countries but I only started to learn English when I was 20 years old. I was a high school student with confidence in Mathematics, Physics and Chemistry, but not English. In South Korea, students who learn natural sciences were able to enter highly ranked universities using high scores in Mathematics and two selected science subjects, which is a kind of tactic for college admission. However, I was lazy and my English grade was around 50 to 60 percentile in South Korea, while Mathematics and Chemistry were always around 70 percentile – a decent score. However, the tactic that I used for collage admission was so risky and it was a disaster. I missed by 2 points out of 50 in SAT Chemistry by mistake and due to this, my grade scale for Chemistry was dropped by one level and 5 universities out of the 6 that I applied for were gone. At that time, I was depressed and blamed the Korean SAT system. Furthermore, it was difficult to get a job in Korea due to the high competition. Even if I could a job, there was no liberty – workers often struggled with night duties and extra work resulting in substantial over time. After I heard about it at the age of 20, I decided to learn English to go abroad for university education and career. Firstly, I went to Philippines to study English and to earn a certificate in English, The International English Language Testing System (IELTS); which is required for overseas university admission. However, it was harsh for me since I had not learned much about English. I could not speak nor write in English. Most of the time, I could not even comprehend when someone spoke to me in English. What I could do was only interpret easy sentences or paragraphs, such as the reading section of usual English tests. There was an academy for IELTS in Philippines and I took an entrance examination which has the same format to the actual IELTS, but my result was bad as I had expected. I got only 3.0 point while the score that I need for admission was 6.0. It was a large gap and seemed rather impossible to increase the score from 3 to 6 in just 3 months. It was April when I entered the academy and new term for university was in August. If I could not obtain the score in 3 months, I had to wait until the next term or join the English course, which was conducted by Management Development Institute in Singapore (MDIS). However, I thought it was a kind of wasting time.

LIFE BEFORE SINGAPORE

At that time when I started to learn English, I was deliberating what country is better for me to go for my studies. Australia, United Kingdom and Singapore were within consideration as being in an English-speaking country would be very helpful. Furthermore, I would have two choices after I finished studies – one was to stay in foreign country to work and the other was to come back to South Korea. If I return to Korea, an ability in English is considered a guaranteed qualification as well as a career. In the meantime, Singapore is a popular country among Koreans due to its high level of security and education. Furthermore, it would be a good challenge for me to learn both English and Chinese. Many news reported Singapore as economically and environmentally developed with high gross domestic product (GDP). It is known as the financial and business hub of Asia. This was very attractive to me and an important reason for choosing Singapore for my university life. As looked forward to the new life, I concentrated most of my time brushing up my English skill for the next 3 to 4 months.

THE FIRST YEAR IN SINGAPORE

My first arrival in Singapore was at the end of July 2015. At Changi Airport, I was impressed with the facilities and pulchritude of the architecture. After that, for one week before the orientation in MDIS started, I travelled around Singapore. There were many fantastic attractions, gorgeous night view and places that showing culture of various countries. Time passed very quickly during enjoyment, as Anthony G. Oettinger said, *"Time flies like an arrow"*. I noticed that the English pronunciation in Singapore was very different to what I had learned in Philippines. During orientation, I was embarrassed as I could not totally understand what was being said. Nevertheless, the diploma course in MDIS was quite straightforward because I had learned most of the content taught in the course in Korea.

"Life is like riding a bicycle. To keep your balance, you must keep moving"

-- Albert Einstein (Buck, 2016)

I am a taciturn person and most people of this kind are slow to voice themselves. During the diploma course, I managed to comprehend spoken English but I still lacked conversational ability. I was still the same person who could just solve questions written in English tests. There were around 60 students in same course. I was able to make some friends and it was good thing that age gap was not an issue because in South Korea, we should have formality even if someone is only a year older than me. Instead, we could study, do assignments and play together. I thought that when I had finished the diploma course, there would be enough time to relax and enjoy my hobbies. After I finished the diploma course, I chose to major in Biotechnology for my undergraduate while all friends whom I had made chose the other programme – Biomedical Science. There were only around 15 students in the Biotechnology class. Starting with first year, we had to take modules which I had never heard of. These were entirely new content to me. Totally unawareness of technical and scientific terms made me struggle a lot and I did not have anyone to ask. Large parts of the lectures were not understandable and I had to look up many things online to help me. If someone wanted to come to Singapore or go to other country for studies, I would recommend them to read and study specialty publications (such as science-related books) and scientific terms to prevent being tossed into an ocean of doubts. If not, studying would be just memorization without understanding and such knowledge would be forgotten easily. The most important thing is confidence. No man is born wise; therefore, if there is any doubt about study, it is necessary not to be afraid and ask someone.

I had learnt many things, from searching for information and scientific writing skills to applied biotechnological skills and knowledge. Laboratory sessions were quite unforgettable, inter alia, Polymerase Chain Reaction (PCR) which is widely used in molecular biology to replicate specific DNA segment. Notwithstanding, the most memorable year of study was the last year.

THE HONOURS YEAR IN MDIS

The honours year was most busy year for me, maybe for everyone as well. There were several demanding modules such as the Final Year Project (FYP) and Investigative Microbiology (IMY). Since I have failed a module previously and needed to repeat the module, I was out of sync with my starting group. My introvert and inactive personality had brought me many disadvantageous. Information about anything related to my studies was always lacking. I did not even know that I had to apply for my FYP topic in March. So, I remembered the day that I went to office to apply for an FYP topic in late April, but the deadline had already passed. It made me very frustrated until I met Dr Maurice HT Ling. He accepted me to join his FYP group and I started my project later than other students. My allocated project was related to the Bioinformatics and Research Methods for Applied Science module which I took in second year. As compared to other students, my project was entirely computer-based which has no laboratory constraint in place. I learnt basics of Python programming and its usage, scientific hypothesis, verification of hypothesis with statistical approach and comparison of proteomic sequences. My project was on comparison of sequence properties of specific archaebacteria and finding correlations to explain archaebacterial biology. At first, I was just wandering and had no grasp of the project; however, I was getting the hang of it with much help from Dr Maurice. If he had not guided me intensely, I was not likely to finish the project. He suggested solutions and tried to fix whenever I made a mistake and that was very helpful. Even though he was my supervisor, his attitude was more like a friend and that gave me confidence. Once, my laptop broke down and I had lost all my files. At that time, I got nervous and embarrassed because they were the files consisting of almost 3 months of work. At that time, I went blank and I did not know what I should say to Dr Maurice. Nevertheless, under his calm attitude and rational advice, I was able to resume the work with full of confidence, and it was one of the reasons why I respected him.

Besides the FYP, the last year was the period that perhaps I most frequently visited the website of National Center for Biotechnology Information (NCBI) to find research papers, nucleotide sequence and so on. In addition, I really appreciated the students that were grouped with me for group presentations and microbiology project. Almost every day except on weekends, we went to laboratory for IMY project, measuring the results week after week. The topic was the effect of temperature to growth rate of 3 yeast species and juice spoilage. Through measuring the properties of fruit juice containing 3 yeast species (*Saccharomyces cerevisiae*, *Schizosaccharomyces pombe and Zygosaccharomyces bailli*), we got to know how the yeasts affected the juice depend on temperature. It was also an interesting and memorable project because all of us did responsibly. It was a good chance for me to get a better understanding on using experimental equipment. We suffered many experimental setbacks but finally able to get good results. The results showed that juice spoilage was not only related to temperature but also related to the yeast species, because some yeasts remained active in low or high temperature. Through the Bioethics module, I was very satisfied with having logical and organized thought. It was helpful that the lecturer encouraged me to discuss about my opinion. With acquisition of applied biotechnological knowledge, the last year was worthwhile.

MY REFLECTION

I do not think that the university life was enjoyable, but it is undeniable fact that the period was both memorable and unforgettable, as what Paul Theroux said *"Travel is only glamorous in retrospect"* (Hendrickson, 1979). I realize that there was no easy way to achieve something but before I came to Singapore, I was always looking for an easy way. Ideologies of every people are not going to be the same and even my views had changed over the years and will changed again. There is no right answer for our life, but it change to fit into our ideals. I do not regret to be a more passionate person; however, I still think that there was a small change in me during my last year in MDIS.

> *"True life is lived when tiny changes occur"*
> -- Leo Tolstoy (Ritchie, 2017)

> *"They must often change who would be constant in happiness or wisdom"*
> -- Confucius

Looking back, these proverbs were what I had realized over the last 4 years. Studying Biotechnology in Singapore gave me much confidence and I appreciate the changes in my mindset and habits. Over these 4 years, there were many instances of self-questioning, like why I did not do well, why I was poor at time management. Nevertheless, I believe that it is a process of growing up, like Andy W. mentioned, *"Isn't life a series of images that change as they repeat themselves?"* (Bockris, 1989), or what William Ellery Channing once said, *"error is the discipline through which we advance"*. If I have stayed in my country and not in the new environment where I had to get accustomed to people and culture or at a place where I do not need to struggle; I would never change.

Honestly, since the first year in Singapore, I did not think that I can endure the environment and pursuing my undergraduate degree because it was much harder for me as compared to other educational programmes that I had taken so far. However, nothing is impossible. Even though I am not superior nor an honours student, I was able to safely finish 4 whole years, like a former American President John F. Kennedy once said, *"Our problems are man-made; therefore, they may be solved by man. And man can be as big as he wants. No problem of human destiny is beyond human beings"*. (Kennedy, 1963).

For these 4 years, I appreciate my family for they always believe me and above all things, I appreciate my parents for giving me a chance to study in Singapore. Life is short, what I will do must be chosen by me. As of now, I am still considering about life after graduation. I can either take a master course, get a job, or go back to Singapore once again. I only live once; hence, the choice should be done very deliberately because it will decide my way of life. As of now, as a person who have national service ahead, which is compulsory for a South Korean male, I cannot work nor study. However, I will have enough time to think about my life after this. But if I ever have chance to be able to work in Singapore, I will like to come back to Singapore.

If anyone hope to study abroad in Singapore, or other countries, I will completely recommend it because I am fully satisfied with the knowledge that I learnt in MDIS about Biotechnology with good teachers around. Even though I did not manage to learn Chinese as I am not an industrious student, I believe that early birds can seize the opportunity. Staying in the same place is same as frogs that never know about outside world. Even if two fishes are the same species, it can grow bigger in the ocean than the one living in fish bowl. *"Life is either a daring adventure or nothing"* (Keller, 1956). Most people who are considering about studying abroad are usually young. In my opinion, it will be a good chance to challenge or to make a turning point in their life. Although I regretted that I did not accomplish what I endeavoured, I never regret about the experiences in foreign country, cultural and study.

REFERENCES

Buck, C. (2016). *"Life is like riding a bicycle; to keep your balance you must keep moving" – Albert Einstein.* Available at: https://www.clairebuck.com/ life- like- riding- bicycle- keep- balance- must- keep-moving-albert-einstein/.

Hendrickson, P. (1979). Paul Theroux, Restless writer of the rails. *Washington Post* (September 20, 1979).

Kennedy, John F. (1963). *Top 10 Commencement Speeches.* Available at: http://content.time.com/time/specials/packages/article/0,28804,1898670_1898671_1898662,00.html.

Keller, H. (1956). *"Life is either a daring adventure or nothing. To keep our faces toward change and behave like free spirits in the presence of fate, is strength undefeatable."*. Available at: https://www.passiton.com/inspirational-quotes/6989-life-is-either-a-daring-adventure-or-nothing.

Morck, R. & Nakamura, M. (2005). A frog in a well knows nothing of the ocean: A history of corporate ownership in Japan. In: *A history of corporate governance around the world: Family business groups to professional managers*, (pp. 367-466). University of Chicago Press.

Ritchie, J. (2017). *Finding your direction: "True life is lived when tiny changes occur" – Leo Tolstoy.* Available at: https://www.lifecoach-directory.org.uk/lifecoach-articles/finding-your-direction-true-life-is-lived-when-tiny-changes-occur-leo-tolstoy.

Victor, Bockris. (1989). *The Life and Death of Andy Warhol.* Bantam Books, New York.

Date Submitted: April 01, 2019. Date Accepted: April 06, 2019.

ABOUT THE EDITOR

Maurice HT Ling
Principal Partner
Colossus Technologies LLP, Singapore

School of BioSciences
The University of Melbourne, Australia

mauriceling@colossus-tech.com
mauriceling@acm.org
computer.in.science@gmail.com

SERIES DESCRIPTION

Current STEM is a broad-spectrum book series targeting practitioners, both academic and industrial practitioners, at all levels of STEM (Science, Technology, Engineering, and Mathematics). This includes all philosophical, theoretical and applied aspects of STEM; as well as STEM-related areas, such as STEM education, STEM industry and economy, ethics and legal aspects of STEM. Hence, Current STEM considers a wide variety of articles as book chapters; this includes, but not limited to:

- Original studies/works (such as research articles, and case studies)
- Updates to previous published studies/works
- Updates or additions to previous methods
- Short communications
- Null results
- Project proposals
- Reviews on research studies / methodologies / tools, etc.
- Algorithm descriptions and software codes
- Educational materials (such as tutorials, and laboratory exercises)
- Dataset communications
- Whitepapers
- Commentaries
- Opinions

In addition to book chapter submissions, Current STEM considers the following:

- Research monographs, such as, theses
- Educational monographs, such as, textbooks
- Collated conference proceedings
- Edited thematic volumes

REFEREE STATEMENT

Unless otherwise specified in the book chapters, every chapter is reviewed by editorial team, supported by reviewers who are either experts or practitioners in the field, before acceptance. In event where a member of the editorial team is an author; the acceptance recommendation(s) of the external reviewer(s), whom is/are not member(s) of the editorial team, take precedence over that of the editorial team. The corresponding author will receive a review report and a letter of acceptance (if accepted) as proof of refereeing.

BOOK CHAPTER SUBMISSIONS

The book chapter be attached (in RTF, DOC, DOCX, or ODF formats) in an email addressed to the Series Editor, Maurice HT Ling, at computer.in.science@gmail.com, with a cover letter stating (1) a summary of the current submission and its significance, and (2) contact details of 3 or more potential reviewers (the author may also list unsuitable reviewers).

MONOGRAPHS AND PROCEEDINGS

Research or educational monographs, and collated conference proceedings submissions should be discussed with the Series Editor,

Maurice HT Ling, via email correspondence at computer.in.science@gmail.com, prior to submission as a book proposal is needed for each volume.

EDITED THEMATIC VOLUMES

Current STEM invites experts to serve as Guest Editors for individual edited thematic volumes. An edited thematic volume collection of scholarly or scientific chapters written by different authors collected together in one book by Guest Editor(s). The chapters in an edited volume must be original works (not republished works). As each volume requires a book proposal, potential Guest Editor(s) discuss their intent with the Series Editor, Maurice HT Ling, via email correspondence at computer.in.science@gmail.com.

INDEX

#

2-methylisoborneol, 163, 165, 166

A

academic performance, 46, 63
actinomycetes, 163, 165, 166
airliners, 68, 121
algorithm, 13, 14, 15, 17, 18, 21, 23, 26, 42
antimicrobial, 163
archaebacteria, 172
audit, 68, 72, 73, 95, 96, 98, 99, 100, 102, 107, 108, 113, 114, 115, 121
Australia, 169, 177
autobiography, 155

B

bankruptcy, 156
Bible Studies, 159
big data, 68, 152, 153
bioinformatics, 53, 54, 55, 57, 59, 60, 61, 63, 64, 68, 152, 172
biological and biomedical research, 68
biotechnology, 59, 159, 160, 161, 162, 171, 173, 174, 175

blockchain, 67, 68, 69, 72, 73, 75, 95, 96, 98, 99, 100, 102, 103, 107, 114, 121, 125, 128, 136, 137, 139, 140, 141, 142, 143, 152, 153

C

Central Institute of Mining and Fuel Research, 46
certificate, 82, 161, 162, 169
Changi Airport, 170
chemical, 34, 163
chemistry, 34, 50, 168
classes, 159, 160, 161
cockpit voice recorder, 68, 121
coding, 1, 2, 3, 5, 9, 31
command-line user interface, 69
comprehension, 2, 20, 21, 22, 25, 26, 32
computational biology, 53, 57, 59, 62
computer, 3, 5, 53, 56, 68, 158, 172, 177, 180, 181
cram education, 168
credit delinquent, 156
curriculum, 46, 49, 53, 64, 159, 164
cyanobacteria, 163, 166

D

data authenticity, 67
data management, 67, 152
database, 69, 72, 112, 116, 117, 118, 119, 123, 124, 125, 126, 128, 129, 130, 131, 133, 142, 146, 147, 148, 149
dormitory, 168

E

EAFP, 31, 32, 35
education, 46, 50, 55, 56, 58, 59, 60, 61, 62, 63, 64, 168, 170, 179
elementary school, 157, 168
emotion, 34, 40
emotional intelligence, 49, 62
English, 2, 60, 157, 158, 160, 168, 169, 170, 171
enumerate, 2, 17, 18, 19, 20, 22
examinations, 47, 48, 159, 162
exception, 2, 30, 31, 32
experiences, vii, 45, 50, 55, 58, 155, 164, 166, 175

F

financial, 51, 56, 58, 156, 170
flight data recorder, 68, 121
format operator %, 10, 11
formula, 24, 158

G

General Public License version 3, 67, 68, 75, 115, 120, 122, 132, 144, 150
generator expressions (GE), 2, 32, 33, 35
geosmin, 163, 165, 166
Germany, 164
grades, 51, 57, 162

grandmother, 46, 156, 157
gross domestic product, 170

H

high school, 49, 55, 156, 159, 160, 168
honours, 45, 50, 52, 56, 161, 162, 167, 172, 174
honours year, 56, 172

I

idiomatic Python, v, 1, 2, 3, 4, 5, 8, 9, 16, 18, 20, 21, 33, 34, 39, 40, 41
idiomatic Python 3, 41
idiomatic swap, 2, 12, 13, 14, 15, 36
if-statement, 32
immutability, 67
indexing, 28, 29, 30, 31
India, 45, 46, 49, 50, 55, 57, 61, 62
Indian education, 55
Indian National Scientific Documentation Centre, 46
insertion, 77, 78, 82, 125, 126, 128, 141, 142, 143
investment, 31, 64, 156

J

Java, 10, 40, 41, 42
JSON, 8, 35

K

kindergarten, 156

L

lambda, 2, 23, 24, 25, 26, 27
languages, 9, 27, 31

Index

learning, 2, 16, 50, 54, 57, 168
list comprehension, 2, 15, 20, 21, 22, 25, 32, 33
living, 155, 175
loan shark, 156
logging, 73, 74, 75, 94, 128, 142

M

Management Development Institute in Singapore, 169
Management Development Institute of Singapore, 45, 49, 155, 159, 167
managing adversity, 51
map, 23, 24, 25, 36
mapping, 72, 98, 99
mathematics, vii, 47, 50, 53, 157, 158, 160, 168, 179
matplotlib, 35, 41
mentor, 45, 54, 56, 58, 62, 63
mentor-mentee relationship, 54, 56, 58
middle school, 157, 168
modules, 50, 52, 160, 161, 171, 172

N

nameless function, 24, 26
national service, 175
Northumbria University, 45, 50, 155, 160, 161, 163, 166, 167
notary, 68, 69, 70, 71, 72, 74, 76, 84, 85, 86, 87, 88, 89, 90, 91, 92, 106, 107, 108, 109, 111, 112, 113, 114, 115, 116, 117, 118, 119, 135, 137, 144, 146, 147, 148, 149, 150, 151

O

Olympiad, 168
organic chemistry, 46

organic compounds, 163
ORM, 7, 35
ORM design pattern, 7

P

parents, 46, 49, 56, 156, 157, 164, 168, 174
PEP, 5, 6, 9, 32, 35, 40, 41, 42, 43
PEP 20, 5, 6, 40, 42
PEP 289, 32
PEP 8, 9, 41, 43
Philippines, 155, 158, 159, 161, 164, 169, 170
physics, 11, 50, 168
platform, 71, 72, 75, 79, 80, 81, 86, 116, 117, 123, 146, 151
Poland, 46
polymerase chain reaction, 171
postgraduate studies, 57
pride, 2, 8, 34, 42
printf, 10
profit, 75, 115, 120, 122, 132, 144, 150
programming, 2, 4, 5, 8, 21, 27, 34, 42, 53, 56, 172
project, 2, 34, 45, 53, 54, 56, 63, 163, 164, 167, 172, 173
proverb, 168
psychological well-being, 64, 65
psychology, 2, 53, 59, 60, 64
psychology of programming, 2
Python 2, 3, 41
Python 3, 3, 43, 70
Python programming, 5, 8, 34, 41, 172
Pythonic, 9, 20, 27, 33, 34, 43
Pythonist, 15, 16, 17, 18, 20
Pythonista, 2, 16, 40

R

random number, 13, 14, 16, 18, 19, 39
reading, 3, 4, 9, 25, 39, 50, 169

recall, 17, 50, 53, 55
reflection, v, 45, 155, 167, 173
reminder operator %, 10
research project, 45, 53, 56, 61, 63
research records, 67, 68, 121
researchers, vii, 161, 163
reversed, 2, 21, 22, 23
rotating-credit association, 156
running time, 14, 38, 39

S

Saccharomyces cerevisiae, 173
SAT, 169
Schizosaccharomyces pombe, 173
school, 46, 49, 55, 156, 157, 159, 160, 168
schooling, 47, 62
science, iv, vii, 1, 46, 47, 49, 50, 51, 53, 60, 61, 62, 64, 152, 159, 160, 161, 165, 166, 168, 171, 172, 177, 179, 180, 181
scientific inquiry, 54, 55, 57
selection sort, 13, 14, 15, 16, 35
Singapore, ii, v, 45, 49, 50, 51, 55, 56, 57, 59, 61, 62, 63, 65, 67, 155, 160, 162, 163, 164, 166, 167, 168, 169, 170, 171, 173, 174, 175, 177
Singaporean education, 50
slicing, 2, 22, 27, 28, 29
software, 2, 9, 34, 76, 116, 120, 122, 132, 144, 150, 179
South Korea, 155, 156, 157, 158, 163, 165, 167, 168, 170, 171, 175
sprintf, 10
SQLAlchemy, 7, 8, 42
stress, 46, 58, 62, 63
string formatter, 2, 9, 10, 11

studying, v, 33, 46, 49, 50, 56, 58, 60, 62, 155, 157, 159, 167, 171, 174, 175
swap, 4, 12, 14, 22, 37

T

Tagaytay City, 158, 159
teachers, 159, 175
textbooks, 34, 55, 180
tuple, 12, 17, 18, 22, 73

U

undergraduate studies, 46, 49
United Kingdom, 168, 169

V

variables, 4, 9, 12, 46
verbose, 1, 16, 17, 22, 23, 39

W

web, 69, 74, 85, 112, 117, 118, 119, 144
web service, 74, 117, 144

Y

yeast, 173

Z

Zen, 4, 5, 6, 7, 8, 40, 42, 43
Zygosaccharomyces bailli, 173

Related Nova Publications

Current STEM. Volume 1

EDITOR: Maurice H. T. Ling

SERIES: Current STEM

BOOK DESCRIPTION: As a book series, Current STEM aims to be a friendly forum for both academic researchers and industrial practitioners to present their work as book chapters. Hence, the chapters should be varied, and this is intended. Current STEM encompasses the type of work to encourage a generation of researcher-practitioners.

HARDCOVER ISBN: 978-1-53613-416-2
RETAIL PRICE: $160

Enhancing STEM Motivation through Citizen Science Programs

EDITORS: Suzanne E. Hiller, PhD and Anastasia Kitsantas

SERIES: Education in a Competitive and Globalizing World

BOOK DESCRIPTION: The chapters describe effective components of citizen science programs, curricula guidelines for K-12 and post-secondary courses, research findings on the impact of citizen science programs for student self-motivational beliefs, achievement, and STEM motivation, as well as guidelines for evaluating citizen science programs.

HARDCOVER ISBN: 978-1-53616-038-3
RETAIL PRICE: $230

To see a complete list of Nova publications, please visit our website at www.novapublishers.com